Working-Class Environmentalism

Karen Bell

Working-Class Environmentalism

An Agenda for a Just and Fair Transition to Sustainability

Karen Bell
University of the West England
Bristol, UK

ISBN 978-3-030-29518-9   ISBN 978-3-030-29519-6   (eBook)
https://doi.org/10.1007/978-3-030-29519-6

This Palgrave Macmillan imprint is published by the registered company Springer Nature Switzerland AG.
The registered company address is: Gewerbestrasse 11, 6330 Cham, Switzerland

This book is dedicated to the victims and survivors of environmental classism
'Remember the dead, fight for the living' International Workers Memorial Day slogan

Acknowledgements

I would like to thank the following people:

The team at Palgrave Macmillan, especially Beth Farrow and Tamsine O'Riordan, for their enthusiastic response to my book proposal and their dedicated work to make it a reality.

Those who commented on sections of the book—Ian Gough, Ian Robinson and anonymous reviewers. I followed their advice on a great deal but have remained stubborn on a few points. Therefore, all mistakes and foolishness are my own doing (or undoing).

My fantastic neighbours, Dave (RIP) and Joyce, who cheerfully help everyone they can.

All my friends around the world, especially those who have helped me to get this book completed with moral, and sometimes practical, support: Laura Dickinson, Lynn-Marie Sardinha, Rita Cangialosi, Lani Parker, Elena Mayorga, Hannah White, Clare Paine, Julie Boston, Jafar Sahari and Jane Bowen.

Lindsay Wall who understood the need for, and encouraged me to write, this book.

My family—including the Bells, the Stanleys, the Bagleys and the Leedhams—for their humanness, thoughtfulness, good humour and caring, especially Jenny, Andrew, John, Philip, Armando, Doreen, Naryano, Jtindriya, Micky, Rachel, Anna, Dorothy, Fred, William, Maria, Ashley, Shane, Chelsea and Emily.

My colleagues in the University of Bristol and University of West of England, especially those who have inspired me, supported my work or with whom I have discussed the ideas in this book. I am particularly grateful to the support staff, the cleaners, administrators and technical staff, mostly working-class people, without whom my academic work would be impossible.

The people whose interviews appear in this book. They are an inspiration with their honesty, humour, caring and strength. I am really grateful for their time, thoughts and words. I hope their messages will be listened to and acted upon.

I also need to thank the UK Economic and Social Research Council for awarding me a Future Research Leaders grant to investigate fair and equitable transitions to sustainability. This gave me the time and security to embark on this book.

Finally, I really appreciate and thank all the authors referenced in this book, whose work I have built on and whose ideas have inspired me, helped me understand and given me hope.

Contents

1

Introduction: Environmental Classism

As I began to write this book, a fire took hold of Grenfell Tower, a 24-storey tower block containing social housing flats in Kensington, London. For years the residents had complained about health and safety issues and had even predicted the possibility of a fire. But they were ignored. The flats had recently been refurbished with cladding, supposedly to help with insulation and prevent damp, though some suggest its purpose was solely to improve the look of the building for the benefit of those living in the surrounding luxury properties. Whatever the reason, this very cladding proved to be the tinder for the fire so that it spread unusually rapidly. Evidence given to the subsequent, and currently ongoing, Grenfell Tower Inquiry (2018) indicated that cost-cutting on the refurbishment, and failing to install sprinklers and other safety features, was a significant factor underlying the ensuing tragedy. The local fire service had also seen drastic cuts and outsourcing to private companies possibly causing the reported lack of basic firefighting equipment at the scene of the fire (Booth 2018a, b). In response to this evidence, one of the members of the survivors' group, Grenfell United, said, 'loved ones would be alive today if different decisions had been taken and if the people in charge had put safety first' (Ruiz 2018).

© The Author(s) 2020
K. Bell, *Working-Class Environmentalism*,
https://Doi.org/10.1007/978-3-030-29519-6_1

I believe there was an important factor underlying all of these mistakes and oversights—'classism'—the majority of the residents of the block were working-class. The casual approach to their safety, the apparent undervaluing of their lives and the dismissal of their concerns could occur because the Grenfell residents did not have social and economic power. Grenfell made classism visible in the horrifying sight of working-class people being burnt alive. It is tragic and should never happen again. But, in terms of working-class lives lost and warnings ignored, it has happened again—and every day since. In the UK and around the world, working-class people are killed and injured through living and working in toxic and dangerous environments every day, largely invisibly, out of public sight and awareness. When they voice their complaints and concerns, they are ignored and sometimes insulted. Because these deaths and injuries in part result from insufficiently valuing working-class lives and concerns, I consider it to be 'classism' and, because it is linked to environmental quality, I call it 'environmental classism'. Though sometimes less visible, direct or immediate than the classism evident at Grenfell, it can be just as lethal and devastating.

Environmental classism, while rarely articulated as a concept, is not a new story. As far back as 1845, Engels described widespread injuries and deaths among the working-class in England due to unsafe factory and housing conditions. He called these 'social murder' (Engels 1845, p. 26), meaning that they were deaths caused by society, rather than by individual intent. Even before then, in 1713, Bernardo Ramazzini wrote *The Diseases of Workers*, the first comprehensive presentation of occupational diseases maiming and killing the working-class in Italy. Ramazzini outlined the health hazards resulting from chemicals, dust and metals faced by workers in more than 50 occupations at that time. Working-class people, themselves, have often been at the forefront of raising awareness about these issues. For example, in 1924, in the US, young women and girls who were working at painting watch dials with a liquid containing radium and mesothorium raised concerns about occupational toxicity. They had been losing their teeth, becoming sick and disfigured and, eventually, dying from bone cancers. Their illnesses were dismissed as being due to a myriad of other causes, including poor dental hygiene, syphilis and even 'hysteria', until the condition was eventually diagnosed as 'radium jaw' and as resulting from their work. The US Radium Corporation attempted to suppress this information. Eventually, five of the young women workers sued the company. One of them died during the trial but, finally, in 1928, the company were forced to pay compensation and the episode resulted in a change in policy, so that scientists, rather than manufacturers, came to determine the occupational hazards posed by radium dial painting (see Abrams 2001, for this and similar working-class environmental histories).

In all these cases, working-class voices had been ignored with tragic consequences. This book is an attempt to amplify their voices and honour their lives. It describes and illustrates 'environmental classism' in the UK and beyond for the purposes of understanding how to end it. 'Environmental classism' refers to policies or practices that impact less favourably on working-class individuals and groups with respect to the quality of their living, working and leisure environments. The book explains how working-class people tend to carry the environmental burdens for society; how they are sometimes negatively impacted by environmental policy and alienated by traditional forms of environmentalism; and how they have long been, and continue to be, environmentalists, even if they have not been recognised as such.

I came to an acknowledgement and understanding of environmental classism as a result of over 30 years of campaigning as an environmental activist and more than 10 years as an environmental justice researcher. Throughout this time, I have heard many stories about the harm that has been done to working-class people as a consequence of their toxic, unsafe and inadequate environments. When I was conducting the research for this book, I heard even more of these stories, sometimes recounted to me in quite a matter-of-fact way. For example, one of the women I spoke to, who lives in an area of high deprivation and near a number of waste facilities, said:

> I really like it and I have a beautiful garden. I really love the place and I got a very beautiful view. I can see … all the mountains if the sky is clear. But sometimes I think I may want to move because of the pollution in the area … There's

quite a lot of pollution and I heard lots of people are ill with cancer. I am one of them, actually. (Ange)

I am writing this book to provide support and solidarity for those who are aware of these problems and to provoke questions and debate among those who currently are not. The book is not meant to be a comprehensive overview of the topic but rather a personal analysis and an invitation to further consideration.

Is Class Still Relevant?

The meaning and relevance of class is contentious. This book draws upon sociological theories of class which include socio-cultural distinctions, such as status, education, intergenerational advantages, and political and social capital which afford influence in the public sphere (see Chap. 2 for debates and definitions).

It has been alleged that class is no longer relevant because of the combined processes of deindustrialisation in the Western world and the rise of neoliberalism globally. Deindustrialisation led to a shift from traditional manual labour to clerical and service sector work, with an increase in management and professional roles and a widening of participation in higher education. At the same time, the rise of neoliberalism led to an increase in 'flexible' employment, with job insecurity increasingly being the experience of all those in the workforce. As a result of these changes, some analysts went as far as to pronounce 'The Death of Class' (Pakulski and Waters 1996). Others argued that class might still exist but, as a mega-theory, was of less relevance in the new social and economic landscape (e.g. Giddens 1991; Beck 1992; Baumann 2001). It is also commonly said that class is particularly peculiar to Britain or the UK.

The social distinction that is more often drawn is between 'poor' and 'non-poor' or 'deprived' and 'non-deprived' socio-economic groups. However, though there is a good deal of overlap, being working-class is not always synonymous with having a low income (Savage et al. 2013;

Savage 2015). Some people may be excluded and disadvantaged as a result of social practices and cultural issues and not just material deprivation (see Chap. 2).

Until very recently social class was rarely discussed in relation to environmental issues or mentioned in mainstream environmental or sustainability reports and studies. Similarly, again until recently, working-class organisations, such as trade unions, had been largely ignored by environmental academics; as well as by the mainstream environmental movement and environmental policy makers (Räthzel and Uzzell 2011, 2013). Some social policy analysts have written about environmental issues in relation to poverty (e.g. Fitzpatrick 2011, 2014a, b), but rarely with regard to class. In the academic fields of 'environmental justice', 'climate justice' and 'energy justice' and among associated activists, the focus has mainly been on race and/or low income, though there have been intermittent allegations of 'classism' directed at the mainstream environmental movement (Sandler and Pezzullo 2007).

However, class is becoming increasingly difficult to ignore. Several authors are now highlighting the extent and impact of inequality in our societies, for example, Danny Dorling (2015), Stephen Wilkinson and Kate Pickett (2009). Some have focused directly on 'classism' as a social injustice (e.g. Jones 2011; McKenzie 2015; Hanley 2016; Isenberg 2016). Classism also became more visible following the UK Brexit vote in 2016 as noted, for example, in *The Independent*, which carried a headline stating: 'Classist innuendo about educated Remain voters and the "white van men" of Leave has revealed something very distasteful about Britain' (Piercy in the Independent, 20th June 2016). As Chap. 2 describes, and the rest of this book illustrates, class is very much still with us. It is not only to be found in the past or solely in England. As a number of sociologists have noted, 'social class, based on ranking, hierarchies and inequalities are found nearly everywhere' (e.g. Silva 2015, p. 373).

It is important to think about how to end environmental classism and embrace working-class environmentalism, firstly, so that working-class people do not have to be killed or injured unnecessarily and, secondly, because it will enable us to achieve sustainability, or a habitable planet for all humans in the long term, as the next sections outline.

Reducing the Health Impacts of Inadequate Environments

Our living and working environments are very influential in determining the extent to which we are healthy or sick and, therefore, the length and quality of our lives. It is often difficult to 'prove' the environmental determinants of health outcomes because of a lack of detailed or appropriate monitoring; long latency times; and the complexity of measuring cumulative, multiple and synergistic exposure to toxins. However, the evidence, taken overall, shows that there is a major link between the environment and health. Environmental factors can degrade health, both directly, by exposing people to harmful agents, or indirectly, by disrupting the ecosystems that sustain life.

The prevalence of many non-communicable diseases is rising globally, with environmental factors being one of the main contributors (WHO 2018a). In the UK, recent research has shown that a century-long rise in life expectancy has stalled since 2010 before the natural human lifetime limit (i.e. biologically, we could live longer and do in some other countries) (Marmot 2017). In another recent study, Hiam et al. (2017) found that UK mortality rates were increasing. Although, in both cases, the authors of the reports speculate that these outcomes are a result of inadequate spending on health and social care under Conservative and Liberal Democrat austerity regimes, an additional explanation may be the environmental assaults on their health that people are increasingly exposed to. The health and social care services mainly step in after we have become unwell or disabled, so we need to think about why we are becoming ill in the first place. For example, though UK cancer survival rates are at a record high, we are much more likely to get cancer. There is currently about a 50% chance of each person getting cancer in their lifetime (Cancer Research UK 2018). Why? We are repeatedly told that the reasons for the national and global increase in this disease are our lifestyles, our genes, our increasingly long lives and more intensive diagnosing, or even bad luck. These reasons may be accurate in some cases but they require further scrutiny. For example, though there is some evidence that cancer rates have increased because people are living longer, it is an inadequate overall explanation because there has also been an escalation in age-adjusted incidence rates (e.g. Pellegriti et al. 2013; SEER 2013). Increased longevity does not explain the rise in childhood cancers.

There is a strong link between poor environments and cancer, as well as many other non-communicable diseases. The Lancet Commission on Pollution and Health (2017) recently estimated that global pollution, alone, kills nine million people prematurely a year and 'threatens the continuing survival of human societies' (p. 465). In the most severely affected countries, pollution is responsible for more than one in four deaths. The study reported that the most health-impacting pollutants were air pollution, linked to heart disease, stroke, lung cancer and other illnesses; and water pollution, associated with 1.8 million deaths per year as a result of gastrointestinal diseases and parasitic infections. Workplace pollution, including exposure to carcinogens, resulted in at least 800,000 deaths. A recent WHO global assessment (2016) similarly highlighted the 'devastating impact of environmental hazards and risks on global health' (p. viii), discussing more than 100 diseases and injuries related to the quality of the environment. Their analysis showed that 23% of global deaths (and 26% of deaths among children under five) are due to environmental factors. Strokes, ischaemic heart disease, diarrhoea and cancers were the main related illnesses. Yet another recent study confirmed these findings again, reporting that there are 71,000 deaths a week globally as a result of outdoor air pollution alone (FIA 2017). Numerous other studies over the last decade have found similar links between poor quality environments and a range of illnesses including heart disease (e.g. Silverman et al. 2010); cancer (e.g. Grant 2009); strokes (e.g. Kettunen et al. 2007); pneumonia (e.g. Knox 2008); respiratory illness (e.g. Knox 2008); adverse birth outcomes (e.g. Shah and Balkhair 2011); breathing problems (Goeminne et al. 2018); impairment of cognitive development (e.g. Ranft et al. 2009); diabetes (e.g. Raaschou-Nielsen et al. 2013); depression and suicide (e.g. Kim et al. 2010); and impeded brain development in children (e.g. Grandjean 2013). Though low-income countries are currently the worst affected, middle-income and high-income countries are also strongly impacted.

Although estimates vary, it is considered that about 50,000 people die in the UK prematurely each year from respiratory, cardiovascular and other illnesses associated with outdoor air pollutants, alone, such as NO_2, particulate matter (PM) and ozone. Outdoor air pollution kills more people in the UK than second-hand smoking or road traffic accidents (COMEAP 2010), yet, until very recently, it has received scant attention. In the city where I live, Bristol, crowned the Green Capital of Europe in 2015, a recent report estimated that 300 premature deaths occur each year as a result of air pollution—about 8.5% of the annual deaths (Laxen et al. 2017).

Therefore, numerous studies now point to the importance of the environment for our health though, even these, may underestimate the burden of disease and death attributable to environmental causes, as only a fraction of the potential risks have been adequately investigated.

Poor environments undermine the health of all of us but the vast majority of those who die and are injured in this way are poor and marginalised (Lancet Commission 2017, section 3). Because of the link between environmental quality and health outcomes, inequitable environments result in increased health inequalities. The least well-off experience worse health than their wealthier counterparts and these differences manifest themselves from the neighbourhood up to the global level (CSDH 2008). In the UK, those living in the poorest neighbourhoods die, on average, seven years earlier than those in the wealthiest areas and spend seventeen more years coping with illness and disability (Ellaway et al. 2012; Marmot 2010; Buck and Maguire 2015). This denotes a double injustice of a shorter life, and a much longer period of poorer health for people living on low incomes.

Although, until recently, few studies have linked environmental, income and health inequalities, those that have generally endorsed the view that a driver for geographical differences in health is the disparity between health-promoting or health-damaging aspects of the environment. For example, almost 20 years ago, the 1998 Acheson report summarised over 700 studies on health inequalities in the UK related to a range of environmental and social conditions, including housing quality, fuel poverty, transport and food poverty (Acheson et al. 1998). Again, in 2004, the Sustainable Development Research Network (SDRN), a former advisor to the UK Department for Environment, Food and Rural Affairs (Defra), reported that environmental inequality can affect peoples' health outcomes (Defra 2004). More recently, Pearce et al. (2010) found that, in the UK 'multiple environmental deprivation increased as the degree of income deprivation rose. Area-level health progressively worsened as the multiple environmental deprivation increased' (p. 522). At the global level, health is undermined because many people still do not have their environmental needs met. For example, approximately 60% of people globally were still without access to safe sanitation systems in 2015 (WHO & UNICEF 2017); 15% lacked access to electricity (World Bank 2017); 30% were without safe drinking water (WHO & UNICEF 2017) and 11% had insufficient food to meet the minimum daily energy requirement (FAO 2015). These studies emphasise the importance of the environment in shaping health and illustrate the "triple jeopardy' of social, health and environmental inequalities' (Pearce et al. 2010, p. 522).

There is international legislation stating that everyone has the right to live in a world free from toxic pollution and environmental degradation (see Boyle 2012), but these rights are not being realised. The poor quality of our environments is disproportionately killing working-class people and, consequently, traumatising their families and friends. But the causes of the resulting disease and death are presented in individualised and fatalistic terms. The role of the environment, until very recently, has often been minimised or ignored. As the Lancet Commission (2017) Executive Summary stated, 'For decades, pollution and its harmful effects on people's health, the environment, and the planet have been neglected both by Governments and the international development agenda' (p. 1). I have witnessed this in academic conferences over the years. For example, in 2012 at a UK social policy conference I attended, a former academic and government health policy advisor claimed the environment influenced only 3% of health outcomes, way below that of the impact of personal behaviour choices. Five years later, the same level of denial prevailed at a conference of 300 researchers, policy makers, business leaders and civil society practitioners gathered to address global sustainability challenges. Most were saying that they did not know how to engage the general public in environmentalism, so I suggested that we need to tell people about the environmental impacts on our health, because people do not seem to be aware that living in toxic environments is behind many illnesses and deaths including the higher incidences of cancer. The Chair of the meeting immediately responded by insisting that the reasons for the increase in cancer were lifestyle, genes and longer lives. If poor quality environments were accepted as the reason for many of our health problems, public anger could no longer be avoided and the situation would, eventually, have to change.

Achieving Sustainability

Now I will turn to the second reason that it is important to end environmental classism and embrace working-class environmentalism: Because it will enable us to achieve sustainability, or a habitable planet for all humans in the long term. While human deaths are occurring now as a result of environmental destruction, we are also on a trajectory towards even greater global devastation. As a result of the overuse and misuse of resources, we have now either crossed or will soon be about to cross, nine earth system 'planetary boundaries' beyond which impacts will be irreversible and the survival of humanity will be unlikely (Steffen et al. 2015). In just one of these boundaries, climate change, we've already increased the earth's temperature more than 1°C and, this alone, has led to numerous disasters, such as floods, hurricanes, heat-waves, droughts and famines. But, if we continue on the current trajectory, we can expect widespread devastation. The 2018 IPCC report stated that we must reduce carbon emissions to net zero by 2050 to have a reasonable chance of limiting global warming to 1.5°C. If we do not now change the trajectory of the rate of greenhouse gas emissions, we may have only 12 years before we irreversibly overstep planetary boundaries (Intergovernmental Panel on Climate Change 2018). McKibben (2017) and many other analysts now argue that we are unlikely to avoid 2°C, no matter what we do. However, the Intergovernmental Panel on Climate Change (IPCC) report (2018) states that we are still in a position to avert the worst scenarios though it will require rapid, far-reaching and unprecedented changes in all aspects of society.

We should also consider that 'climate change' is just one of the nine planetary boundaries. Another that is in danger of transgression is 'biosphere integrity', including loss of biodiversity. For example, a recently published study of insect life in Germany found a decline of 76–82% in flying insects (Hallmann et al. 2017). This is highly problematic because, as one scientist stated: 'Insects make up about two-thirds of all life on Earth but there has been some kind of horrific decline … We appear to be making vast tracts of land inhospitable to most forms of life, and are currently on course for ecological Armageddon. If we lose the insects then everything is going to collapse' (Goulson 2017, np). We rely upon insects for the pollinating which is necessary for our food supply, so there will be widespread hunger if insect populations collapse. 'Ocean acidification' is another planetary boundary that may soon be overstepped and may lead to the loss of many species of fish and other sea life (Bioacid 2017). Added to this, the United Nations Environment Programme (UNEP) has recently warned that, at current rates, 'By 2050, there could be more plastic than fish in our oceans' (2017, np), a trend also seriously impacting on marine life. Whilst being devastating events in themselves, the impact on humans will be tragic. International scientists from the Intergovernmental Science-Policy Platform on Biodiversity and Ecosystem Services (IPBES 2018) are now predicting that the unsustainable exploitation of the natural world could severely threaten the food and water security of billions of people. Some have more starkly described the scenario in terms of the near-certainty of 'civilizational collapse' (Erlich 2018).

Therefore, a rapid and effective transition to sustainability is required. This requires more than technical innovation; it also needs structural and institutional change, supported by a transformation in the beliefs, attitudes and behaviour of individuals. Such a change depends upon there being a broad base of people who are willing and able to advocate for environmental and social justice, as well as to make personal choices that will benefit the environment. This broad base has not developed to the extent that might be desirable or necessary, given the urgency and seriousness of the environmental problems that we now face.

This book argues and evidences that the development of a healthy environment for all and the base for environmental transition has, fundamentally, been constrained by divisions between different social groups, in particular social class divisions. Although there are other factors driving unsustainability, including the dominant economic system, social mobilisation could challenge these factors. I argue in this book that social mobilisation has been undermined by 'environmental classism' and I encourage environmentalists to better integrate a working-class perspective into their environmentalism. Working-class people bring particular expertise to the debates and actions. In class societies, the wealthier are less likely to be impacted by the environmental crises because they can change their dwelling places, jobs and other aspects of their lives in the face of risks and crises. Without such options, working-class people may be the first to be impacted by these crises so have the most to gain from avoiding them. In addition, those who have no vested interest in the current structures of power can find it easier to see when the system is not working (Krauss 1993) and may, therefore, be more likely to identify what needs to be addressed and changed. A revived and supported working-class environmentalism could, therefore, drive forward a rapid, fair and effective transition to sustainability.

Achieving this will require removing the barriers to working-class environmentalism. This book will identify, explain, and make recommendations on, how to overcome the barriers between working-class groups and the mainstream environmental movement, environmental practitioners and environmental policy makers. In this context, the 'barriers' referred to mean the attitudes, beliefs, policies, practices and cultures that prevent effective cooperation.

The situation seems to be reaching a 'tipping point' where environmental issues are now increasingly being recognised as fundamental to human health and wellbeing; trade unions are beginning to grapple with ecological issues; and new coalitions for social change are emerging between environmentalists and social justice campaigners. This seems a crucial time to look at how bridges can be developed between environmentalists and diverse working-class communities so as to ensure that sustainability strategies are more attuned to social and cultural difference. I am hoping that this book will help build these bridges by discussing how class is experienced in relation to the environment through the everyday realities of life.

What Is 'the Environment?'

Before going further, it is important to ask this fundamental question. As with the notion of 'class', there are also debates and contentions about the concepts of 'environment' and 'environmentalism'. In a classic text, Agyeman (2002) noted the tendency for the mainstream UK environmental movement to use a narrow definition of the term 'environment', one defined by, and of relevance to, White, middle-class people. In the US, academics have contrasted the predominantly middle-class mainstream environmental movement with the more working-class environmental justice movement in terms of implicit assumption about what denotes 'environment' and is the focus of an environmental agenda. For example, Allen et al. (2007) contrasts the middle-class environmentalist, 'engaged in bird watching, recycling, "buying green," hiking', with the working-class environmentalist concerned with 'children playing in the shadow of smokestacks', noting 'what the environment is (fragile ecosystem to be protected versus a place of dangerous threat)' differs dramatically (Allen et al. 2007, p. 124).

The working-class-based 'environmental justice' movement usually focuses on issues that relate to human health, while the mainstream environmental movement has often focused on preserving wilderness and the protection of endangered species (Bullard and Wright 1992; Di Chiro 1996). The environmentalism of environmental justice activists addresses the real and perceived threats to health where we live, work and play. Pulido calls this an 'environmentalism of everyday life' (1998, p. 30). Therefore, when working-class groups approach mainstream organisations for support with their issues, for example the location of incinerators in their communities, the mainstream organisations can deem such issues to be outside the scope of 'environmental'. For example, Di Chiro describes an environmental organisation refusing to support a community in addressing an environmental threat because 'the poisoning of an urban community by an incineration facility was a "community health issue," not an environmental one' (Di Chiro 1996, p. 299). Behind this may have been a fear that concerns for animals and wilderness would be given low priority in a movement which also focused on anthropocentric issues (Allen et al. 2007). However, the capture of the term 'environmentalism' by middle-class people has meant that working-class people are less likely to relate to or to identify with the movement. Hence, definitions invite inclusion and exclusion and facilitate certain alliances whilst undermining others, so that 'the environmental movement has shot itself in the foot by adopting the definitional frontiers that delegate different issues as either inside or outside the environmental "frame"' (Di Chiro 1996, p. 279).

Synergising these positions, in this book, I use the widest possible understanding of the term 'environmentalism' incorporating the 'live, work and play' as well as the 'wilderness and species' aspects. I believe all are important and need to be integrated to achieve equitable and fair environmental transition.

In some ways, it does not matter if we approach environmentalism from a desire to protect other species or from a more anthropocentric, human-focused, motivation. So long as we are not ignoring or discounting the other approach, since that only divides the movement and increases the risk of developing inadequate and partial solutions. Some have argued that, though there is overlap, attaining objectives in one dimension may be insufficient for another and objectives may even sometimes conflict (e.g. Dobson 2007). There is a lack of empirical research on this question, but it does not seem to me that there should be any inherent contradiction. It would certainly be possible to improve some aspects of human environmental justice at the expense of other species, for example, as with industrial agriculture which may improve short-term agricultural yields, yet disrupt ecosystems and, thereby, habitats for other species. However, I would argue that, whilst alleviating some problems, such practices would, ultimately, also be detrimental to humans who have also evolved within ecosystems. Environmental health seems to be a precondition for human health.

Method, Scope and Structure

The book builds on previous research from the fields of sociology, social policy, environmental studies, geography and politics, as indicated in the references throughout. It particularly draws on the academic literature relating to environmental justice, climate justice and energy justice. I also use my own experience and personal anecdotes, as consistent with the demands of the environmental justice movement to recognise the importance of storytelling as an epistemology, equal to that of scientific and elite knowledge bases (Krauss 1993). Much of my analysis focuses on the UK, in particular council estates, deprived inner-city areas and pockets of semi-rural poverty. However, the book also makes links internationally and considers global processes and implications, referring to low-, middle- and high-income countries.

Initially, a comprehensive search strategy was developed to locate all the relevant published evidence to date. Subsequently, I looked at my own research on environmental justice over the last ten years (2006–2016), reanalysing some of the qualitative elements from a class perspective including that published in Bell (2008), Bell and Sweeting (2013) and Bell (2014). I also carried out 27 targeted interviews specifically for this book using informant-directed interview techniques. All the interviewees were working-class according to at least one definition (see Chap. 2) and, importantly, all but one were brought up in working-class homes with parents who undertook non-professional work and had had no further education. They have all been given pseudonyms to protect their identities. Their main occupational identities are (or were, if retired or currently unemployed): taxi-driver (Ray); gardener (Pete); security guard (Dave); builder (Jim); bus driver (Liam); academic (Mel); housing worker (Sharon); receptionist (Sheila); warehouse workers (Janet, Anne, Joan); technical workers (Paul, Carole); care workers (Cath, Julie); caterer (Ange); shop or other service workers (Stacey, Pat, Cindy); hairdresser (Kelly); foster carer (Amy); maritime worker (Bob); community worker (Anita); nurse (Jo); aircraft engineer (Jack); factory worker (Phil) and administrator (Mick). Very few work in industries that would traditionally be seen to be harmful to the environment and this would be an important area for future research since such workers may have a different perspective. However, their occupations do roughly reflect the current breakdown of the workforce in the UK—that is, with four out of five working in the service sector and few in manufacturing or energy (Resolution Foundation 2019).

The interviewees' ages range from early 20s to late 80s. Seven identi-fied as Black, Asian and Minority Ethnic (BAME) and 20 as White. They live in cities (Bristol, Manchester, Reading), towns (Lowestoft, Aldershot) and rural areas of England and Wales. Most of those interviewed live on council estates or in inner-city areas and almost all the interviewees live in areas which are in the 10% most deprived areas of England and Wales.

I used the following recognised strategies to decide who to include in the research: Those who had particularly relevant knowledge and experi-ence ('purposive sampling', see Mason 2002); those who I could access via networks ('snowball sampling', see Gilbert 2001); those who lived in the maximum diversity of environmental situations; and those who fitted a range of demographic descriptors in terms of age, gender, LGBTQ+, disability, ethnicity and so on (i.e. 'maximum variation sampling' see Miles and Huberman 1994).

The interviews were semi-structured, in that I asked a number of key questions each time, but the additional questions varied enormously between interviews and often took the form of a conversation. The ques-tions included: Would you call yourself working-class? What do you like and dislike about the area that you live in? Would you say it was a healthy place to live or not? Have you ever tried to improve the local environ-ment? Have you ever joined an environmental campaign group or thought about it? Have you met any members of environmental organ-isations? and so on. Any vague or contentious terms were clarified during the interview, as necessary.

As well as the interview aspect of the research, I included knowledge gained in relation to my environmental justice activism over the same period, which had been recorded in notes, emails, letters, webpages and blogs. This included my activities and involvement in meetings, events and activities related to political parties, local government, community events, campaigns and activist groups. The personal stories that I draw on include, for example, those relating to my experience as an environmen-tal campaigner and of living on council estates throughout my life. During this time, I witnessed many struggles for working-class people to be heard by environmental policy makers and environmental activists.

The identities of individuals interviewed or referred to in the context of the situations described have been anonymised to avoid any unwanted publicity. Prior to the interview, the interviewees were always fully informed as to the nature of the study and gave their consent to associated anonymised publication. Where I have referred to public events in which it was not possible to gain full informed consent from all those present, I have also anonymised comments. It seems reasonable and ethical to protect the people present from any harm or unwanted attention that could arise from repeating their words outside of the fora where they were originally articulated. The only exception to this is where I occasionally refer to those in public office making public comments in their official capacity.

In relation to the reflections on activities I was personally involved in as an activist, it was not possible to gain informed consent retrospectively. I have, therefore, also anonymised the individuals and situations concerned, so that they are not identifiable. However, in a few cases, I could not anonymise the situation completely without losing important information, for example, in relation to Bristol Green Capital (since Bristol is the only city that has ever been designated a 'Green Capital' in the UK to date) and my involvement with the Green Party (since there is no other primarily environmental party in the UK). In these cases, I have just anonymised the individuals concerned.

My work is rooted in a 'Critical Theory' framework, where research is 'engaged' in that it seeks to have a positive influence on society (Pacheco-Vega and Parizeau 2018). As a scholar-activist, it was not possible to be 'value free', 'neutral' or completely 'objective' but rather I aimed to be 'reflexive', that is, aware of my subjective position at all times whilst attempting to see beyond it (Kincheloe and Mclaren 2000). To be engaged and to also meet scholarly standards of credibility, I must acknowledge my own assumptions, values and ideologies. Therefore, a reflexive and engaged approach requires considering 'positionality'—that is, to think about ourselves in the research and how we influence it; and how we can interact meaningfully in the activities related to the research without being exploitative and causing harm. The last section of this chapter refers to my background which should aid understanding of my positionality. Also, I need to state outright that I am a social and environmental justice campaigner, and have alternated between membership and activism within the Green Party and Labour Party since the 1980s. I am now currently in the Labour Party.

I do not see my task as to provide a balanced view of environmentalism in relation to working-class people, but rather to offer insight and understanding about working-class views of environmentalism, whilst at the same time being thoroughly honest so that these insights might be of value. I feel that this is legitimate as there is already so much that has been written about environmentalism from a middle-class perspective. However, to ensure ethicality in the research, I was careful to consider the possibility of other interpretations; to be scrupulously honest about my observations; and to make clear distinctions between observations, interpretations and their intersections. I also carried out thorough documentation of the research process; collected as much data as possible; used a variety of methodologies; integrated the research with the existing literature and checked my interpretations during the interviews and informal conversations, as much as possible.

Even so, I could never claim that this study is a complete reflection of reality. I would need far greater resources and time to achieve this. I have only captured a partial range of viewpoints and experiences, albeit rooted in a lifetime of experience of these issues. However, because of the lack of other literature on this topic, I have no definite way of knowing the prevalence of these experiences and views in the population or how much they reflect reality. I do not claim to present 'the truth' but rather to shed some light on the situation. My ambition, therefore, in the chapters that follow is to translate, explain and analyse environmentalism from a working-class frame of reference in order to offer some possibilities about how to bring about greater sustainability and environmental justice.

When I speak of the 'working-class', I do not mean only the White working-class but all working-class people. It is often argued that class-based analysis fails to sufficiently explore the interconnections between class and other forms of oppression, for example, on the basis of gender, disability, race and LGBTQ+ (e.g. Rubin et al. 2014). In order to take into account the additional or different impacts of other oppressions, we speak of 'intersectionality' which emphasises that class cannot be experienced outside of other identities. It is clear that race, disability, LGBTQ+ identity, age and other intersectional considerations will also shape outcomes for working-class people and these are discussed, when relevant.

One of the BAME working-class women that I spoke to for this research, Mel, was very keen to make this point, stating:

Class and race is very intertwined … You know, there's an unrepresentative amount of BAME individuals who are pushed to the lowest class of the society so more and more, we're talking about …a more diverse group than the general population and, I think, whether you're White working-class or Black working-class, your voice just gets marginalised, the things that you care about most will get marginalised. (Mel)

Intersectional factors, such as disability and BAME status, often compound the environmental injustices that some working-class people experience since, in many instances, they also go hand in hand with poverty and unequal status. Where once class was seen as the primary social division in society, as a result of feminist, anti-colonial and other emancipatory struggles, there are now ongoing debates about the relationship of class to these other potential sources of disadvantage. It is, therefore, important to note that using a framework that focuses solely on any single issue will not be sufficient to capture the form and extent of injustice that most working-class people experience. Though I focus on class in this book, I refer to the other identities where relevant. This does not mean that I consider class to be the most important identity in determining everything in our lives, but rather that I believe it has been relatively overlooked, particularly in relation to the topic of environmentalism.

Chapter 2 explains more on the definitions and nature of class and classism, including sections on intersectionality, microaggressions, discrimination, internalised oppression, the impacts of classism and how classism is justified. Chapter 3 details how working-class people in the UK are more likely to experience environmental deprivation, for example, in terms of air pollution; transport; proximity to landfill sites; flood risk; food poverty; fuel poverty and access to green space. Chapter 4 explains how environmental improvement programmes, environmental transition processes and programmes—locally, nationally and internationally—have often excluded working-class people and low-income groups, sometimes compounding their disadvantage. This theme is explored in relation to some key environmental issues from street cleanliness to energy justice. Chapter 5 discusses how working-class and other disadvantaged groups are often left out of environmental decision-making processes, despite the important knowledge and perspectives they bring. Chapter 6 outlines and analyses the key contributions of working-class people to environmentalism through a discussion of the 'environmentalism of the poor' (Martinez-Alier 2003); the environmental justice movement; and trade union health and safety campaigns. It challenges a prevailing view of environmentalism as a 'post-materialist' struggle (e.g. Inglehart 1990) which suggests that poorer people are too preoccupied with meeting their basic needs for food, warmth and security to be able to think about, or be active on, environmental issues. Chapter 7 discusses the reasons for environmental classism in terms of a number of theories. Chapter 8, which concludes the book, discusses the theoretical, policy and practice implications of this analysis.

The key aim of this book, then, is to amplify working-class voices on environmental classism. I am quite anxious about doing this, as I fear (1) letting down working-class people by not doing it well enough; (2) further stereotyping working-class people and stigmatising their communities; and (3) offending my many middle-class environmentalist friends and colleagues.

Regarding the first, I am very conscious to be fair and honest as outlined above. I have also asked some of the participants in the interviews to read through drafts of this book and/or discussed with them how to represent the issues that they raised.

In relation to the second, while attempting to capture relevant patterns or generalities about working-class lives, I reflect on a comment in the book, 'White Working-class' (Williams 2017). Joan Williams, the middle-class author, recounts the time when, as a young woman, she went to meet the family of her new working-class boyfriend. Upon hearing that his family complained that 'she looked at us like a fucking anthropologist', she admitted, 'I was cut to the quick … because it was so true' (Williams 2017, p. 5). Though, unlike her, I am from a working-class background, I want to avoid doing the same, and I particularly want to avoid unwittingly reinforcing stereotypes. So it is important to emphasise that making generalisations does not imply that all working-class people have the outlooks and experiences described here. There is a rich and complex variety of lived experience among working-class people and all of it must be respected if we are to end environmental classism. It is important to emphasise that the stories are not necessarily remarkable, but they are ours—mine and those of the other working-class people interviewed. I hope the quotes convey the sense of this book being about real people and not just abstractions. The personal is, indeed, the political, and it is through the everyday actions of ordinary individuals that classism is most obviously felt. I also do not want to further stigmatise council estates and inner-city areas. They are often fantastic places to live because of the wonderful people you find there, and some of these areas have some very positive environmental aspects too, such as there being an abundance of green space on the outer estates. But it is important to be honest about what needs to change.

With regard to the third concern, I just want to reassure my middle-class friends, and other middle-class or owning-class people, that I do not intend to lay the blame for this situation at the door of individuals, even though I believe that as individuals we can do much to improve it. I think we are all guilty of making mistakes on equalities issues, even when we are members of oppressed groups ourselves. It is a product of our socialisation within capitalist, patriarchal, racist, disablist and homophobic societies. In many cases, individuals are prevented from behaving as we would ideally wish by the political and economic system. Our thoughts and behaviours are always, to some extent, constrained by this context. This is not to let anyone off the hook, but it is important to contextualise the problem. All we can do is try to become aware, to learn and to change our thoughts and actions whenever possible and to join together to try to change the context. Chapter 8 outlines the many things that we could do to make the changes needed to reduce, and then eliminate, environmental classism and support the further development of working-class environmentalism.

About Me

In order to illustrate the complexity of working-class life, and to further position myself in the research (i.e. let you know what drives and influences me), I want you to know something about my background. As you will see, my experience can fit all the theoretical positions on how to define working-class, as detailed in Chap. 2, in terms of exploitation, lower status, lower incomes, lack of power, intersectionality and lack of recognition. However, I do not represent all or any other working-class people. I also wish to highlight that, despite a number of 'accomplishments' usually associated with being middle-class—a PhD, an academic job, many publications and a mortgage, I still consider myself to be working-class because the path to get here was much longer and less smooth than for most middle-class people. I do, however, recognise that there are many reasons for life being a bumpy ride, and that middle-class people also suffer (perhaps even more in some ways). The first reaction from almost all the middle-class acquaintances that I have spoken to about writing this book is 'but you can't be working-class if you are writing a book/have a PhD/work at a University', as if everything I accomplish that they consider to be not for the likes of us suddenly takes away my class background. This seems quite unfair when they know almost nothing about me, so I hope the following short life story also explains why I still identify as working-class.

My childhood experience of being working-class is fairly average for the time (1960s/1970s), somewhere between the extreme marginalisation described by Lisa Mckenzie (in *Estates, Class and Culture in Austerity Britain* 2015) and the more respectable, somewhat less marginalised, existence described by Lyndsey Hanley (in *Respectable* 2016). I come from a manual working-class family (with some Romany gypsy heritage) and have lived almost all my life on council estates, including now. My sister and I were brought up by my maternal grandparents (here referred to as 'Mum' and 'Dad'). Formerly both domestic servants, throughout my childhood my Dad worked as a boiler stoker in an army camp whilst my Mum, who was not altogether well, stayed at home. To try to make ends meet, my Dad worked a number of other jobs on top of his state job, usually involving gardening for wealthy people. This meant that he was not at home much. I have many memories of my Mum crying because she could not work out how to pay for everything each week and of her, my sister and I hiding under the table because she could not afford to pay whoever was knocking at the door demanding settlement of rent, utilities and credit. Poverty undermined her mental health and so the house was turbulent to say the least but survivable because there was also a lot of laughter, fun and affection. We received some food and clothing from extended family relatives and my Dad's workmates. My Mum and Dad also helped other people, cooking food and making clothes for relatives and helping neighbours with their gardens and any emergencies. All the helping happened very informally and unsentimentally. Fortunately, we lived in a council house which meant we did not pay for repairs and we had a degree of housing security, as long as we could pay the rent. My Mum, who had always grown up with the threat of the workhouse, was very careful to pay the rent before anything else. Our house was always very well cared for and we were all well dressed, everything bought via the mail order catalogue and hire purchase. Very little in the house was ever replaced. The house, and everything in it, was cared for on the basis that it must last forever. It could never be broken or damaged as there was simply no money to replace it. This was another source of stress for all of us, as there were so many rules necessary to make everything last as long as possible. When I took a middle-class friend to my family home later on, she said 'your parents are so ecological!' She was seemingly interpreting poverty and thrift as environmentalism. The 40-year-old tray that our tea was served on had almost all the former paint worn off, with the bare tin showing though, for example. Still functioning, it was not a priority to replace. I don't think my parents knew the word 'ecological' but, it was true, poverty had made them more ecological than many environmentalists.

Our garden was well tended and my Mum and Dad modelled how to appreciate nature, in the form of their care for the birds, plants and life that lived and visited there. Like nearly every other working-class family at the time, we did not travel abroad, though we had a few holidays by the sea staying in a caravan with the wider family. I remember the first day back at school after the summer holidays, the teacher would always ask who had been abroad. Nearly all the hands shot up but then the teacher would say 'not including the Isle of Wight' and nearly every hand would go down again.

I estimate that there were less than ten books in the house. I discovered the local library about a year before I left school but at that time it was very small and not very inspiring. However, we did have the *Daily Mirror* to read every day which then had some excellent journalists, such as John Pilger. Also, my Dad, as a Labour Party supporter and trade unionist, constantly talked about politics and current affairs.

In 1970, I passed the 11+ exam which gave me the opportunity to go to grammar school. My mum did not want me to go since she could not afford to buy the uniform—I usually wore hand-me-down school uniform. My junior school teacher (Mrs Murrell) wrote to my mum to encourage her to let me go. My Mum appeared honoured to receive a letter from a teacher but could still not afford the uniform. The situation was resolved when, shortly after, the grammar school system was abolished, so I was sent instead to a comprehensive, socially mixed school.

Since my family struggled to make ends meet, I started work aged 12 collecting charity donations and then collecting shoes door-to-door for repair. I did these jobs for three hours an evening after school each day, as well as most weekends. When, aged 15, I saw a sign for a job in a clothes shop on the way home from school, I successfully applied for it and never went back. I then worked in a number of manual jobs, including as a factory worker and a migrant worker in Europe. Because of the poverty and other difficulties at home, I felt it important to leave as soon as I could so that I would no longer be a burden. I started off working in hotels in the UK where I could 'live in'. I then began to go further afield for work, to Europe, spending several years working on farms and in tourism, often sleeping on building sites, in phone boxes or under trees to save money because of the low and insecure wages. During one period of unemployment as an agricultural worker in Spain (the crop was late), like many other seasonal workers, I sold my blood plasma to a blood-bank for £5, a sandwich and a glass of milk. From this, I contracted a painful, life-threatening illness which almost killed me and undermined my health and wellbeing for the next 35 years. I have now recovered from the illness but, of course, I can never make up for the havoc it wreaked on my life. When people tell me I am no longer working-class because I have a PhD, or have written a book, I want to tell them, 'yes, but when I was unemployed, hungry and homeless, I could not ring anyone up to send me money, I had to sell my blood' but I don't say this because I feel ashamed and I wonder if they would believe me.

After almost a decade of manual and casual employment, thanks to the encouragement of an environmental science teacher and a now long-gone, UK maintenance grant-funded university education system, I was eventually able to go to college and then to polytechnic. I studied a range of earth and social science subjects and, later, I took a qualification in youth and community work. I then went on to work as a community development worker in deprived areas of Bristol on issues of environment, participation, health and equalities. About ten years ago, I began working on a PhD on environmental justice and then hoped to embark on an academic career. Academic work is notoriously hard to get, especially for working-class people who do not have the necessary contacts and confidence. So I worked on very short-term, multiple, casual, academic contracts until 2016, sometimes so under-employed that I had to rent out my own bed and sleep on a floor to make ends meet. As a result of this very stressful experience at my former workplace, I developed a supposedly permanent, stress-related neurological condition which meant I lost my voice almost completely apart from an incomprehensible croak. However, I persevered with my academic work and, over the course of several years and using various natural methods, my speaking voice has gradually almost returned to normal. I now have quite a secure contract and work with very supportive and inspiring colleagues at the University of the West of England.

I am tempted to edit out some of this life story because I feel some shame about mistakes I made, even though they were the result of situations that were not my fault. I think this is a typical reaction to trauma and that is why working-class stories remain hidden much of the time. I am also tempted to edit out some parts of the story because they do not seem to relate to environmentalism. Again, I believe this illustrates some of the thinking that has got us to where we are, separating social from environmental justice. It is all thoroughly connected, and this book will highlight how. What this brief life story does show, however, is how utterly dependent we are on social and environmental policy—for decent housing, education, jobs, health, sanity and happiness. It also, perhaps, explains why I have bothered to write this book. I know that, if I had not been working-class and seen all that I have seen, I would probably not be so desperate that people do not have to suffer with illnesses, insecurity and deprivation. This drive, alongside the fact that we are living through, perhaps, the most critical period in the history of humanity with, perhaps, just 20 years to resolve the environmental crises, has motivated me to try to make a difference. I hope this book will also motivate others to resolve these crises and alleviate this unnecessary suffering. While presenting a limited series of interviews and participatory experiences may not be enough to turn around the situation, I hope it will raise questions. I hope the book will lead to shifts in thinking and that the discussion will extend beyond academia and elites. Environmental classism, like any other classism, means that the needs of the majority of society are ignored and, sometimes, deliberately. But it can, and it must, end.

2

Class and Classism

Before focusing specifically on 'environmental classism' it seems important to say more about how class and classism manifest and impact on
our lives more generally and to justify the way I use the concept of 'class'
in this book. The main indicators for class that I use are: personal and
parental income, occupation and education. I will briefly outline the
issues and the rationale for my focus, but I will not weigh up the different
definitions of class in depth because they frequently overlap and the surrounding debates are fundamentally irreconcilable.

Theories

The concept of social class has developed through the work of a number
of theorists including Karl Marx (e.g. 1867); Max Weber (e.g. 1958) and,
more recently, Pierre Bourdieu (e.g. 1984, 1990) and Mike Savage (e.g.
Savage et al. 2013; Savage 2015). As a result, various definitions of class
have been proposed and there have been longstanding and ongoing
debates regarding the most important identifiers and classifications. Class
has been described in terms of exploitation, income, education, status,

© The Author(s) 2020
K. Bell, *Working-Class Environmentalism*,
https://doi.org/10.1007/978-3-030-29519-6_2

occupation, cultural signals, subjective affiliation, lifestyles or some combination of these (for debates, see for example, Wright 2001; Goldthorpe and Chan 2007). In general, though, it has long been agreed that class is multi-dimensional (e.g. Veblen 1899; Weber 1958; Bourdieu 1984) and that members of a class share common 'life chances' which influence the probability that a person can procure goods and find satisfaction (e.g. Weber 1958).

The main divide in the debate about class is about whether someone's class position is best defined in 'gradational' or in 'relational' terms (Ossowski 1963). The everyday use of the term 'class' tends to be gradational, that is, linked to stratification based on people's occupations, educational qualifications, social status and power. This usage is broadly compatible with Max Weber's work. Weber introduced status to an understanding of class, claimed through positive and negative privileges based on (a) style of life, (b) formal education and (c) hereditary or occupational prestige. Weber's position was strongly rooted in an appreciation of class as differential life chances. From Weber's model arose Goldthorpe's class categorisation, which links to income, employment conditions and lifetime earnings (Goldthorpe and McKnight 2006). This operational method was adopted in the UK National Statistics Socio-economic Classification (NS-SeC) and is now used in the UK census, among other surveys. NS-SeC distinguishes eight social classifications based on occupation: Higher managerial and professional occupations; lower managerial and professional occupations; intermediate occupations (clerical, sales, service); small employers and own account workers; lower supervisory and technical occupations; semi-routine occupations; routine occupations; never worked or long-term unemployed.

Many studies followed Weber's work on particular cultural aspects of class. For example, Basil Bernstein (1971, 1973, 1975) wrote about social class in terms of language, arguing that children come to communicate differently according to context—they will speak differently depending on whether they are in family settings, with their peers or in formal education. The family settings may be working-class or middle-class. The latter use an 'elaborated code', as do formal educational establishments. Working-class families, by contrast, use a 'restricted code'. Bernstein saw neither as inherently 'better', but his theory did explain why working-class children may not do as well at school—because their mode of communication was not recognised as equally valuable in the school setting (Bernstein 1990, pp. 118–119). More recent, 'cultural class analysis' (Atkinson 2010) also looks at the differing conditions of life that flow from distinctive class cultures and the dynamics of power and domination that these lead to (e.g. Skeggs 2004; McKenzie 2015). Bourdieu's work has been particularly influential in this regard. His work focuses on the human need and desire to be recognised by other humans and to be seen as worthy and valuable.

A recent study of class, the 'Great British Class Survey', an online survey of nearly nine million people, was based on the assumption that 'classes are not merely economic phenomena but are also profoundly concerned with forms of social reproduction and cultural distinction' (Savage 2015, p. 223). The study identified seven classes: Elite; Established middle-class; Technical middle-class; New affluent workers; Emergent service workers; Traditional working-class and Precariat. Although the project produced some interesting data, numerous doubts have been expressed about this classification and the terminology used. Arguments were made for the validity of other typologies, particularly the National Statistics Socio-Economic Classification (NS-SeC) (e.g. letters to *The Guardian* 5 April 2013 from Guy Standing; Roger Goodger; David Rose and Eric Harrison). Because of the emphasis on lifestyles in the survey, one of the *Guardian* letter writers jokingly suggested a better indicator of class would be shopping habits, so that you would have elite (Waitrose); established middle-class (Marks & Spencer); technical middle-class (Sainsbury's); new affluent workers (Morrisons); traditional working-class (The Co-op); emergent service workers (Tesco) and the precariat (Asda) (The Guardian 2013).

Some have argued that these 'gradational' and 'cultural' definitions miss the wider relational aspects of class in terms of exploitation, domination and unequal divisions of power (e.g. Wright 2005). Class, from a Marxist perspective, reflects relationship to the 'means of production' (the factories, offices, shops and associated equipment), that is, whether one owns the means of production or has to sell their labour power to those who do. Some Marxists hold that society is structured in terms of an owning class, who own most of the means of production and distribution (e.g. factories, offices, farms), representing just 1–2%; a middle-class, of those in command positions (e.g. management and professional) or with small businesses; and a working-class who make up the vast majority of the population. From a Marxist perspective, these classes are opposed to each other and form a hierarchy of power, privilege, authority and financial resources. Due to the limited availability of the command jobs, those with advantage of birth are better able to take up these jobs and deny them to others. Marx stated that 'in so far as millions of families live under economic conditions of existence that separate their mode of life, their interests and their culture from those of other classes, and put them in hostile opposition to the latter, they form a class' (Marx 1852/1972, p. 515). Although Marx mentions culture here, the most important issue for Marx was exploitation. As such, I believe there is an important place for such an analysis when we are discussing the harms caused by capitalism, but the problem for me with using a solely economic Marxist definition at all times is that some variants of Marxism put many high-income professional workers from middle-class families in the category of being working-class (e.g. most lawyers, doctors and MPs). These individuals usually do not own their means of production, but most may have had very different life chances from their working-class counterparts. Marxism might also categorise someone from a middle-class family who has chosen a manual job as 'working-class', even though they could choose a more lucrative or professional job and may well do so later.

Therefore, when I use the term 'working-class' in this book, I am not using a solely economic Marxist conception of the term. My definition includes the material distribution of wealth and income but also 'recognition', status and valuing. In addition, as Lisa McKenzie pointed out in a recent podcast about class, there is an important intergenerational aspect of class (CLASS 2017). Intergenerational occupational mobility, that is, the chance to have a better or worse job than that of your parents, has remained stable in recent decades, or may have declined. A recent report on the outcome of social mobility policy in the UK over the last 20 years showed that our class position is still very much determined by that of our parents (Social Mobility Commission 2017) and that the rate of mobility within the UK has declined. The report stated that it will take 120 years before disadvantaged young people are as likely as their better off counterparts to get equivalent qualifications and that the best-paid jobs remain 'deeply elitist'. The same pattern is repeated in many OECD countries (OECD 2018a) with a high likelihood of people being 'stuck' in the income group into which they were born. I believe it is vitally important to consider this intergenerational aspect. Otherwise, as mentioned in Chap. 1, when a person from a working-class background accesses a predominantly middle-class opportunity or chooses to engage in some aspects of a middle-class lifestyle, they could immediately be re-categorised as middle-class, even though their path to that opportunity or choice will have been so different, and usually much more difficult, than that of a middle-class person obtaining the same outcome. To deny our class when we are 'successful' on middle-class terms seems to be to deny our oppression.

The Continuing Importance of Class

In recent decades we have frequently heard that class, by any definition, no longer exists, or is unimportant. We were told, for example, 'The study of class is no longer central to British sociological analysis, and the debate on class is largely about whether this should be celebrated or lamented' (Savage 2000, p. 7). It was alleged that either everyone is now middle-class or everyone is now working-class. We were told that there had been a declining identification with, and general significance of, class as a result of an increase in the number of middle-class jobs, the decline of trade unions, deregulation of the economy, greater social mobility, wider access to education, an increase in precarious work and the break-up of communities (e.g. Beck and Beck-Gernsheim 2002). At the extreme, these arguments characterise society as made up of individuals who have such mixed and complex identities and ways of living that it is impossible to categorise them (e.g. Dalton 2008). Most of those I interviewed identified as working-class, sometimes in spite of discourses that denied their class. For example, Jo and Anita said:

I think I would call myself working-class. My dad was a plumber. My mother was a hairdresser. I grew up on [a] … council estate … I did go to university, entering as a nurse, and I can remember years ago looking at some sort of scale which put a nurse as not working-class but I think nurses are very much working-class. I live on a council estate now and I would say I identify as working-class. (Jo)

I'd say I'm working class because of my childhood, growing up in a single parent family on benefits on a council estate with a wider family who are definitely working-class … my grandad worked in power stations and my aunt worked in factories. So that's my background. Whether that's changed by virtue of an education and what my peer group is now … I might look middle-class but my values and the way I see the world are very much framed from the first 18 years of my life so, yes, I'd say I'm working-class. (Anita)

It has also been argued that new economic and environmental pressures mean that we are all equally under pressure. Beck (1992), for example, in 'Risk Society' considers that risk is now everywhere, impacting on all classes. There have also been assertions that those who engage in informal, insecure employment, whether that be zero-hours contracts or unpaid internships, constitute a new class, 'The Precariat' (Standing 2011).

However, the argument that income and quality of life is converging between the classes is not borne out by the evidence. As this chapter and the book as a whole will evidence, class origins play an important role in shaping life chances. Evans and Tilley, in a recent review of the evidence on class and politics, stated:

Levels of class identification remain robust over more than fifty years … There has been little change in perceptions of class conflict and barriers between classes, or even in the difficulty of having friends from other social classes … Occupational class and education shape the reality of people's lives, in terms of incomes, instability, health, and social mobility as much as they did fifty years ago. (Evans and Tilley 2017, pp. 9–29)

They argue that the perceived disappearance of class was ideological, rather than based on reality, as it suited politicians to ignore or hide the reality of enduring or increasing social divisions, inequalities and injustices. Atkinson's (2015) examination of class in the twenty-first century came to a similar conclusion that 'they may look very different from the past, but class structures, cultures, struggles and modes of domination persist as doggedly as ever' (Atkinson 2015, p. 15).

Many others, whilst not always mentioning class specifically, have pointed to the growing increase in both wealth and income inequalities in the UK and globally (e.g. Piketty 2014; Oxfam 2016; The Equality Trust 2017; Dorling 2018; OECD 2018a, b). In terms of wealth inequality, if you are female, working-class or Black, your pension pot, savings, property, investments will usually be much less than for a male, middle-class, White person, and you are likely not to have any wealth at all. Around 41% of professionals, such as accountants, doctors and lawyers in the UK, have at least £500,000 worth of wealth, compared to just 12% of people in routine occupations, such as shop and service sector workers (IPPR 2017, p. 13). With regard to income inequality, between 1997 and 2017 the gap between the highest and the lowest paid in the UK increased dramatically. In 1998, the highest earners were paid 47 times that of the lowest but, by 2015, the equivalent gap had increased to 128 times (Social Mobility Commission 2017, p. 5). Danny Dorling (2018) reports that the gap between the very rich and the rest is wider in Britain than in any other large country in Europe. In the US, income polarisation has been even more severe (Weeden et al. 2007). Several of those I interviewed for this book made the point that opportunities and quality of life had in some ways deteriorated for the UK working-class in recent decades. For example, Bob, a man in his late 50s, said:

I think we are a lucky generation. In our days the working classes did progress but now we are seeing regression ... Since Maggie ... [Margaret Thatcher]. She destroyed the communities. (Bob)

The UK and US situation is mirrored globally. Estimates suggest that the poorer half of the global population collectively owns less than 1% of global wealth, while the richest 10% of adults own 89% of all wealth (CSRI 2016, p. 12). The 2017 Forbes global rich list reported a record number of new billionaires (Forbes 2017) while one in nine of the world's population goes hungry (FAO 2015). This situation has worsened over the last 40 years as a result of the rise of neoliberal economic policies, globalisation and deindustrialisation in some countries and technological change.

Class difference tends to be rooted in occupation. As well as paying less, working-class jobs tend to be less secure than middle-class jobs and to be associated with less entitlements and 'perks' such as sick pay, holiday pay, pensions, insurances, promotion prospects, flexibility and autonomy. Working-class people experience job insecurity in terms of precariousness but also earnings, with fluctuations depending on hours available, overtime bonuses and so on. A number of studies show that the pay gap between the classes has increased as jobs have become polarised in terms of pay and conditions (e.g. Autor and Dorn 2013; Oesch 2013; Gallie 2015). Statistical evidence from the UK Labour Force Survey of employment rates show that unemployment rates for working-class people have been on, average, three times higher than that of their middle-class counterparts since the 1970s (Evans and Tilley 2017). Changes in the economy disproportionately affect the working-class.

The most important predictor of occupation is education (Breen and Jonsson 2005) and, unsurprisingly, there have been persistent class inequalities in educational attainment (see e.g. Crawford 2014; Stuart et al. 2014). Attending a 'public school' (i.e. a private, fee-paying school) is one of the best predictors of educational attainment, hugely multiplying chances of success. Although only 7% of the UK population attend these schools, their pupils go on to make up 70% of High Court judges and 54% of CEOs of FTSE 100 companies (Sutton Trust 2009). These elites tend to move in their own circles and pass on their privilege to their offspring. Privately educated people tend to marry each other and send their own children to fee-paying schools, thereby perpetuating privilege and class apartheid. As well as poor attainment in school, working-class children are less likely to go on to Further (i.e. post 16, college) and Higher (i.e. post 18, university) Education (Erikson et al. 2005; Bolton 2010) with little change since the 1970s.

Education is important because it is a facilitator for, and predictor of, other life chances. Those who stay longer in education usually go on to earn more over the course of their lives (O'Leary and Sloane 2005; Kirby and Riley 2008; Conlon and Patrignani 2011), with those completing a degree going on to earn about twice that of those who don't (Conlon and Patrignani 2011) and being significantly less likely to experience unemployment (Busch 2015). Four-fifths of students accepted at Oxbridge University between 2010 and 2015 had parents working in top professional and managerial occupations. Who goes to these institutions matters because many of them will eventually run the country. Most of the UK prime ministers, senior judges and senior civil servants have been to these institutions (Richardson 2017).

Class also frequently predicts health outcomes. A number of studies show class differences in terms of health, whether that be based on self-assessments of general health, physiological measures, assessments of health-promoting behaviour or analysis of indicators of mental health (e.g. Drever et al. 2004; Atherton and Power 2007; Roberts et al. 2013; Anderson 2016). There has been a particularly striking rise in the difference between child mental health between the classes over the last 40 years in the UK (Anderson 2016). Death rate differentials between the classes in England and Wales have also increased (Langford and Johnson 2010; Johnson and Al-Hamad 2011). The next chapter discusses how these health inequities may be linked to differences in environmental quality.

Therefore, the evidence and analysis indicate that class still persists and that it predicts life chances in terms of income, occupation, education and health, among a number of factors. The idea that it does not exist or matter appears somewhat ideological or even out of touch. This was also the view of some of the people I interviewed in the course of my research. For example, Dave, who lives in the South of England, said,

I don't know why, but maybe because I live in quite an affluent area of the country, but it [class] seems to be a bit of an outspoken word now. I mean, what is quite clear is that the society is a class society … the people who deny that are usually people in the very privileged position, they're not the people who are at the bottom end of the scale working on low wages, who are struggling to pay their bills, right …. The idea that there's not a working-class I think is, quite frankly in my view, a lot of middle-class crap! It's been something that's been created, you know, to sow the illusion that society has actually moved to one of equality, which it clearly has not, and if anything, it's gone the opposite way. (Dave)

Dave's comment, delivered quite emotionally, also illustrates that denying the existence of class can be upsetting to some working-class people as it seems to invalidate their experience.

Classism

A useful definition of classism and one which I subscribe to for the purposes of this book is the following:

> *The institutional, cultural, and individual set of practices and beliefs that assign differential value to people according to their socioeconomic class; and an economic system that creates excessive inequality and causes basic human needs to go unmet.* (Kirby 2011, p. 141)

All of the above class inequalities could be considered to be the material evidence of 'classism', that is, the marginalisation and unfair treatment of working-class people. Classism includes everyday practices, attitudes, assumptions and behaviours. Like other forms of oppression, it can be institutional or interpersonal. It can be proactively and deliberately enacted or endorsed or it can be passively allowed to continue. In the UK, certain groups are 'protected' in existing equality legislation so that it is not legal to discriminate on the grounds of age, disability, gender reassignment, race, religion or belief, sex, sexual orientation, marriage and civil partnership, pregnancy and maternity. Social class does not have 'protected' status under US, UK or, as far as I know, any other law. As a result, there are no explicit legal or social incentives to address class-based discrimination or prejudice. This also means that there is no mandatory requirement to collect data about class background for monitoring purposes. Therefore, showing evidence of classism can be difficult. In spite of that, there is a huge amount of such evidence, some of which is listed below.

One aspect of classism is 'stigma'. As if the disadvantages of low pay, poor working conditions, inadequate education, greater ill-health and shorter lives were not enough, working-class people are also frequently stigmatised. As McKenzie (2015, p. 7) states, 'although there is an economic dimension to inequality (what you have, what you earn, how and what you are entitled to), there is also a cultural dimension that feeds into stigma (what you wear, how you speak, your accent, your tastes)'. All of these can be used as targets of classism.

Some might argue that our social class can be hidden and, therefore, unlike other forms of oppression, we can avoid stigmatisation and discrimination on the grounds of class. This common view needs to be questioned, firstly because membership of other oppressed groups can also be hidden; for example, some disabled people's impairments are not visible, and, secondly, because there is evidence that class is not as invisible as might be assumed.

A recent study found that people can reliably tell an individual's class (operationalised as monetary income) just by looking at their face (Bjornsdottir and Rule 2017). The same study found that identifying the individual's class enabled them to then use those impressions to discriminate, including when judging who to hire in employment situations (Bjornsdottir and Rule 2017). Prior studies from around the world show that people tend to rate those who appear to be from middle-class backgrounds as being more competent and worthy than those from working-class backgrounds (see Fiske and Markus 2012). In a study by Cozzarelli et al. (2001), for example, participants more often associated negative traits, such as being lazy, stupid, dirty and immoral with poor people than with middle-class people. These prejudices are instilled at an early age—even preschool-aged children demonstrate such tendencies (see Horwitz et al. 2014; Rivera and Tilcsik 2016).

Lott (2002), in a review of such studies, characterised society's classist attitude towards people living on low incomes as 'distancing; that is, separation, exclusion, devaluation, discounting, and designation as "other"' (p. 100). There is some evidence that distancing, when taken to an extreme, can lead to dehumanising. For example, Harris and Fiske (2006) undertook brain scans of study participants being shown photos of people from a range of social groups. When shown photos of rich people, the participant's brain scans showed that the medial pre-frontal cortex was activated. This area is activated when we see another person, rather than an inanimate object. When looking at pictures of poor people, however, this area of the brain did not light up—it was not registering that they were looking at photos of the same species. The poor people seem to have been dehumanised.

Dover's (e.g. 2016) work on oppression and exploitation illustrates the multi-faceted lived experience of dehumanisation. If people are dehumanised they are unlikely to be listened to or considered by policy makers and can more easily be mistreated. McKenzie, in her study of a working-class estate in Nottingham, notes how residents 'described their lives as 'fighting brick walls', no one listening or caring about any of the problems they had' (McKenzie 2015, p. 93). She makes the wider point that working-class people are frequently portrayed as 'ignorant, uneducated and lacking in moral judgement' (McKenzie 2015, p. 105). Beider (2011) similarly notes that working-class communities are often viewed as being problematic and dysfunctional and the individuals within them are ascribed particular attributes, rather than recognising that they have composite identities (Beider 2011). This stigma seems to have increased in recent years. In the UK, Owen Jones expressed this phenomenon as the 'demonization' of the working-class (Jones 2011). The mainstream media, in particular, tends to portray working-class people as promiscuous, drug-abusing, lazy and otherwise dysfunctional, especially those claiming state benefits. The narrative is that our hard-earned taxes are being used for these people to live it up—drinking, drugging and lazing around. McKenzie (2015) argues that politicians, the media and particularly some 'reality TV' programmes leave us with the impression that people who live on council estates lack aspiration, moral values, a work ethic and intelligence.

There have been some studies in relation to people's experience of such classism which found that working-class socio-economic groups feel negatively judged, isolated, humiliated, devalued, looked down on and blamed by those who are more affluent than them (see Simons et al. 2017). Most of the working-class people I spoke to for my research felt that they had been treated less well because of their class. They spoke of being looked down upon by middle-class people, not being taken seriously and treated as if they have no skills, are deficient and stupid, with frequent assumptions of criminality. Sharon and Mel, for example, explained:

Everybody has a skill, everybody has something to offer and I think to assume that poorer people haven't got that skill and, I'm not saying that everybody thinks that, but do you know what I mean, their [middle-class] *way is not the only way … It goes back to that sense of entitlement, arrogance, that their way is the best way … Just the assumptions that people think you're thick … I've been accused of stealing as well.* (Sharon)

Sometimes they hear but they don't listen and there's a difference and sometimes defensiveness can get in the way of actually listening … I have to fight and be stronger in my arguments and more prepared in my arguments because my authority or knowledge will be questioned more, so I feel like I have to justify myself or my stance more … I see, sometimes, [middle-class] *people saying the most stupid things and they're just not questioned because they have that kind of air of authority, you know.* (Mel)

Everyday insults and injuries towards oppressed groups have been called 'microaggressions' (Sue 2010), referring to the sometimes deliberate and sometimes inadvertent, daily indignities, invalidations and slights that are experienced by marginalised groups. Microaggressions can be very explicit verbal or nonverbal putdowns that are intended to hurt, such as name-calling. Or they may be subtle or commonplace exchanges, but they intentionally or unintentionally communicate demeaning messages to people. Microaggressions were originally discussed in relation to racism but the term has recently also been applied to classism. For example, Smith et al. (2015) explored class-based microaggressions in higher education in terms of the messages and treatment that working-class students receive from staff, administrators and peers. The students experienced 'disparaging comments and attitudes' which they found to be 'debilitating' and 'paralyzing' (p. 137). Several MPs from working-class backgrounds have recently spoken out about the classist abuse they have suffered in their role. For example, Ian Lavery, currently the Chair of the Labour Party and a former miner, was criticised for being too loud and having too strong an accent such that the audience could not understand him when he appeared on BBC's Question Time. He responded by saying he would rather 'die in a gutter' than lose his working-class accent (Walker 2017). There was similar criticism of Shadow Education

Secretary, Angela Raynor MP, accused of being stupid on account of her accent. Although these microaggressions may seem trivial as one-off events, as microaggressions accumulate over a lifetime, they can do a great deal of harm, whether intentional or not.

Classist attitudes and beliefs were exposed in the aftermath of the Brexit referendum in 2016. As a result of living around working-class people, I knew that the working-class leave vote was about anger and a desperate need for change, so I was upset when middle-class people, some of them my friends, began to attack the 'leave' voters as stupid and racist. Mckenzie (2017) has written about this in a study of the opinions of working-class leave voters in East London and ex-mining towns in Nottinghamshire. She reports on the hurt her informants felt when they saw the reaction of elites to their vote to leave, accusing them of being 'bigots' and 'backwards'. They had wanted a way to kick back at an establishment that they felt let down by because the communities where their families had lived for generations had been devastated. They did not talk about race or migration but more about 'feeling abandoned, struggling financially and feeling totally remote from mainstream political parties… they just couldn't stand it being the same' (McKenzie 2017, p. 207). Brexit also highlighted how segregated the classes are in UK society, as Sharon, one of the working-class women I interviewed for this book, said:

> *There are many different worlds out there and I'm just thinking about the Brexit vote that there were so many White middle-class people that did not realise that there were people out there that wanted to vote Brexit and that was because they never reach out to that world in any way … assumptions were made that everybody felt the same way.* (Sharon)

Although some academics have challenged the notion that a greater proportion of the working-class voted leave than the middle-class (e.g. Antonucci et al. 2017), their actual data indicates that those with no qualifications or no education beyond GCSE were much more likely to vote leave (educational levels that are much more prevalent among the working-class). Therefore, disagreements about this seem to be based on differing definitions of class. However, whether or not working-class people were more likely to vote leave, accusing those who did of stupidity and racism seems to be rooted in classism.

Classism affects life chances in terms of employment, education and health and undermines confidence and self-esteem. US and UK studies show that employers see those from middle-class backgrounds as more desirable workers (e.g. Jackson 2009a; Rivera 2015). For example, Rivera (2015) found that, when screening CVs, where qualifications and experience were equal, professional service recruiters favoured those applicants who displayed middle-class cultural signals, such as participation in traditionally middle-class sports and hobbies. These employers considered that these signals indicated an ability to fit in with the elite culture and clientele of their company.

Discrimination in education and employment leads to economic difficulties and very stressful lives. Poor people (who will mostly be working-class) face 'elevated rates of threatening and uncontrollable life events, noxious life conditions, marital dissolution, infant mortality, many diseases, violent crime, homicide, accidents, and deaths from all causes' (Belle et al. 2000, p. 1160). Classism may also affect health through physiological stress responses, related to neurochemical, endocrine and immunological functioning, which are associated with both physical and mental health problems (see, for example, Hatzenbuehler et al. 2013; Krieger 2014). Recently, Robert Sapolsky (2018) has discussed this connection, suggesting that relative poverty generates stress which stimulates the secretion of cortisol, a risk factor for mental and physical illness.

Experiences of classism have been significantly associated with feelings of inferiority (Simons et al. 2017). McKenzie (2015) notes: 'They know that they are looked down on, that they are represented as not good enough, and that their tastes are often rubbished or ridiculed. This is hurtful but also has a negative effect on your life chances, your opportunities' (McKenzie 2015, p. 7).

This shame resulting from classism can result in threats to self-esteem through 'internalised oppression'. When people are targeted, discriminated against or oppressed over a period of time, it is not uncommon for them to internalise, that is, come to believe and make part of their self-image, the myths and misinformation that society communicates to them about their group. Sennett and Cobb (1973), in their seminal and influential text, *The Hidden Injuries of Class*, report on their interviews and group discussions with 150 blue-collar workers, revealing how the workers internalised the shame of feeling inferior on the basis of class, undermining their sense of personal dignity. Their interviewees said that they were treated like 'nothing' and like 'dirt' (1973, p. 139). Sometimes this treatment is considered, by those who are at the receiving end of this violence, to be 'the natural order' if it is considered at all. Mel, one of the women I interviewed, made a very explicit comment about this:

> *Because, even though I really fight against it … when I'm in certain environments, I become my own worst enemy. I can question myself. I can—I feel like I don't belong in the room, and almost pull myself out of certain situations because I don't feel like I belong. So, in some ways, it's me doing it to me, even when people are very nice and are not being discriminatory.* (Mel)

I also recently saw a classic example of this 'internalised oppression' with a working-class person believing that she genetically lacked intelligence on a reality TV programme called 'Long Lost Family'. As Davina McCall introduced a working-class mother to the son she had given up for adoption at birth, now a university lecturer with a PhD, she exclaimed: 'I can't believe it—he came out of me and he's got a PhD'. I have yet to meet a working-class person who thinks they are intelligent enough to get a PhD, except those who have one.

Such unquestioned assumptions about the 'natural order' are considered by some social theorists to be manifestations of 'hegemony' (Gramsci 2010). Gramsci used the concept of hegemony to describe a process whereby consent to the continuance of the current structures is manufactured by the powerful through dominant ideologies and values. He drew attention to the power of ideology, values and belief in reproducing class relations. In particular, Gramsci located hegemony in these daily interactions and so-called common sense understandings of the world.

One of the hegemonic beliefs that support classism is the meritocratic ideology that pervades capitalist societies. According to this ideology, being affluent is a result of individual merits, such as having marketable talents and working hard. Those who do less well economically can legitimately be judged as lazy or stupid and less deserving than those who supposedly worked hard to reach the top. This meritocratic discourse allows people to justify, rationalise and make invisible class oppression as well as inculcating a sense of entitlement among the middle and owning classes. It is not surprising, then, that the Joseph Rowntree Foundation (JRF) in the UK found that there was a widespread belief that poverty is the individual's own fault (Hall et al. 2014). In addition, Evans and Tilley's (2017) analysis of data found that all classes thought that the pay rates for particular occupational groups were fair and that people generally believe that individuals themselves, rather than their circumstances, were the reason for success. This corresponds with the findings of numerous other studies showing that people tend to believe there is equality of opportunity and that people get what they deserve.

Other hegemonic beliefs are evident in policy debates that demean poor and working-class people, including the 'underclass', 'problem families' and 'broken Britain' discourses. The role of society in creating social problems is ignored or downplayed and outcomes are explained in terms of individual faults or even genetic make-up. For example, Prime Minister Tony Blair, in his 'problem families' discourse, declared that members of such families could be identified 'virtually as their children are born' (Murphy 2006, p. 2).

The meritocratic ideology is supported by numerous stereotypes about working-class people which can be used to justify unequal outcomes. Working-class cultures are often discussed in terms of being inferior, displaying 'bad taste' or 'incorrect behaviour'. Williams (2017) notes that 'Everything we do is class-marked' (p. 26) and that often what is beloved of the working-class is considered inferior or ridiculous by the middle-classes, giving the example that her circle of middle-class friends would never send a Hallmark card or go to Dunkin' Donuts.

To understand the working-class, the middle and owning classes need to try to grasp, not only working-class culture(s), but also their own culture in terms of its assumptions and truths. One of the problems is that middle-class people rarely get to know working-class people. Research shows most social interactions are with people similar to ourselves—making us less sensitive to how people outside our 'social bubble' live. This can lead to profound misperceptions about the state of society (Windsteiger 2017).

The Three Big Classist Myths

The kind of portrayals of working-class people that we see in the media and in public discourse are often based on distorted, exaggerated classist myths. The most common three are that working-class people are lazy and lack aspirations; that they are racist and intolerant; and that they are self-destructive and anti-social.

The first of these, that working-class people are lazy, is based on the notion that working-class people don't want to work. I call this a myth because evidence shows that, generally, working-class people take up work when it is available even though it may be poor quality (Shildrick et al. 2012). Since working-class jobs are often precarious, involving zero-hours contracts and short-term or insufficient hours and low pay, working-class people are forced into drawing on the welfare system that they also pay into through their taxes when they are working. Many are now stuck in such employment scenarios, in the 'low pay, no pay' cycle (McKenzie 2015). To work hard is a strongly held value among the working-class and most are reluctant to claim the benefits they are eligible for, at least until it is absolutely necessary to survive. For working-class people working hard means 'to do a menial job you hate for 40 years, and reining yourself in so you don't "have an attitude" (i.e. so that you can submit to authority)' (Williams 2017, p. 20). Because this requires enormous drive, discipline and sacrifice, the characteristics working-class people most tend to dislike are dishonesty, irresponsibility and being lazy, as a number of studies have found (e.g. Lamont 2000). Working-class people cannot rely on the second chances that middle-class people can often access, so they have to work hard.

My previous work suggests that working-class young people can be highly ambitious but may be stifled by a lack of opportunities or information about the opportunities that exist or the confidence to take them up (e.g. Bell 2011a). In the research for this book, one of the interviewees said on this topic:

> I can remember being told by one careers advisor to be realistic about, I mean, basically, do you want to work at Tesco or Sainsbury's? There was no talk of college or university. Most people that I know went into vocational training, so

boys would become a mechanic, girls would become hairdressers or you work in a shop or, you know … But I've since gone on and got two Master's Degrees and a PhD, so they were completely wrong about me being realistic. (Mel)

Aspirations may seem like they are lacking in working-class communities because there is a strong emphasis on loyalty to the family and community—not abandoning them to make your own life better. Williams (2017) highlights that working-class people generally aspire to live in their own class milieu but with more money. If you choose to move away for work it could be seen to imply that you care more about your job than your community. Skeggs (1997) highlights how sections of the working-class are encouraged to have aspirations that involve rejection of working-class culture and identity so as to 'improve' themselves. McKenzie describes how working-class people did not generally want to leave their communities but wanted to be successful according to the values of those communities so that 'being "somebody" on the estate is always preferable to being "nobody" on the outside' (McKenzie 2015, p. 98). Sometimes this included having nice cars and nice trainers or just money in your pocket.

The second myth that working-class people are racist and intolerant of minorities is usually directed at the White working-class. This myth is true only to the extent that racism is unfortunately, and tragically, in all White social groups. For example, in a US study of implicit assumptions, about half of the White respondents were found to automatically prefer Whites to Blacks (Morin 2015). But this statistic applies equally to middle-class and working-class US respondents (Mooney 2014). Many White people are racist and that needs to stop urgently, but the answer is not to point the finger at the White working-class. People who are suffering as a result of not being able to get jobs, housing and services do not need to be accused of racism but to be understood as people who are stressed. Some of them, indeed, are looking to blame someone other than their political leaders for their problems, as governments are encouraging them to do. Working-class communities can also be divided along ethnic lines when they are forced to compete for the minimal resources that are available (Taylor and Wilson 2016). They may mock 'political correctness' but more as a critique of the fact that their problems don't seem to matter compared to those who have 'protected characteristics'. As Williams (2017) argues, when elites commit to equality for many different groups but arrogantly dismiss the working-class, this is a recipe for extreme alienation and fury.

Some working-class people can become racist as a way of rationalising their own oppression. However, any racism and intolerance that exists in these sections of the working-class is not congenital, it can be overcome. When working-class people are shown respect and value, they will not need any scapegoats to blame. In his autobiographical first book, *The End of Eddy* (2014), Édouard Louis makes this point. He provides an insider's story of the provincial working-class in France, living on minimum wages and diminishing benefits. He describes the homophobia, racism and violence of his family and community, including towards himself as a gay young person. He considers these aspects of his upbringing as arising from brutalisation and despair and puts the blame on a remote political class that had systematically denied dignity to his family and others like them.

Finally, the third myth, that the working-class are self-destructive and anti-social, also needs to be challenged. Most working-class people know that some of us behave in anti-social and self-harming ways, as Dave discusses in our interview with regard to working-class communities:

> *What I am witnessing is a diminishing of what I would deem people's mental health ... People aren't taking care of things anymore like they used to... They are not respecting the areas where we live. I think, in general, there's a lack of the self-worth that used to exist 30 to 40 years ago.* (Dave)

As Dave's comment proposes, some of this outward destructiveness has arisen in recent years and is underpinned by a lack of self-esteem and wellbeing. Again, it is not a predominant or inherent characteristic of working-class people but has arisen from stressful circumstances following a trend of increasing inequality. Psychologists recognise five basic responses to trauma and ongoing stress—fight, flight, freeze, friend and flop (Levine 1997; Ogden and Minton 2000; Porges 2004). These are methods that have enabled humans to survive over generations. They should be used in the moment of threat but, when there are multiple and ongoing threats, they can become chronic. It might be argued that these correspond to certain irrational behaviours that can sometimes be found among working-class people—anger, resistance and aggression (fight), misuse of drugs and alcohol, other addictions and irresponsible behaviours (flight), apathy and depression (freeze) and strictly complying with social norms—being 'respectable' (friend) and becoming apathetic, submissive and malleable (flop). Moreover, research shows that people have limited amounts of 'will-power' or 'self-discipline' and this gets used up when constantly having to do things that are not what you really want to do (e.g. working somewhere that is boring) so that you can, literally, run out of this willpower (Hagger et al. 2010). Therefore, I consider that some of the self-destructiveness found among working-class people is coping behaviour. Darren McGarvey also makes this point in *Poverty Safari* (2017), while discussing the violence in his working-class home and community, something I also experienced. As he states, 'stress is often the engine room that fuels the lifestyle choices and behaviours that can lead to poor diet, addictions, mental health issues and chronic health conditions' (2017, p. 61).

The sooner that we have equality and fairness, the sooner we can reduce stress and begin to eradicate these problems. Middle-class people also sometimes engage in this kind of behaviour but they are more likely to be able to access resources and be given second chances which enable them to step out of the situations which foster it. This is illustrated, for example, in a recent case of an Oxford University student who stabbed someone but was spared a custodial sentence on the grounds that it could spoil her career (Moore 2017).

The Good Stuff

There is also much that is positive in working-class lives. McKenzie (2015) describes the diversity and richness of working-class communities and cultures, so often misunderstood and demeaned or unseen by outsiders. For example, she highlights the importance of engaging in banter and telling good stories. There is a strong emphasis on being part of and serving the community, as opposed to the valuing of individual accomplishments among the middle-classes. Cohen's (1998) study of mature students from working-class backgrounds found that they chose to define success, not in terms of social mobility, but rather in terms of their ability to serve their communities, become social change agents and advocate for the disenfranchised. In another study, Jensen (2012) identified a working-class preference for belonging vs. becoming. The middle-class tend to strive to become something better, whereas the working-class are concerned to fit in. While elites show their sophistication by pushing boundaries of acceptability in the cultural realm, working-class people often consider it important to maintain and show respect for your own culture so that others do not feel threatened or uncomfortable.

There is a caring attitude towards others in general. A number of studies by Paul Piff suggest that working-class people are more generous, less narcissistic, less entitled and more ethical than their wealthier counterparts. For example, in one study, he gave people £10 each which they could spend on themselves or give some, or all, away. Wealthier people were more likely to spend the money on themselves—those earning less than $25,000 a year gave away 44% more than their wealthier counterparts earning between $150,000 and $200,000 a year did. Yet Piff's work indicates that the generosity has nothing to do with genes or inherent personality traits. In another study, he gave money to those who don't usually have it to see if they would take on the characteristics of the rich. Some were given an advantage in a game of Monopoly—$400 when they went past Go, instead of the $200 given to others. As the 'luckier' group began to win the game they became noisier and took more than their fair share of a snack left for both players. When asked why they thought they had won, they spoke of the effort they had made and their good decisions, not mentioning the advantage they had been given. Piff concluded that, while competitive and selfish people may be good at making money, having money itself can make someone selfish and competitive (Piff et al. 2012). So, again, there is nothing congenital about middle-class or working-class behaviour; it is shaped by living in a class society.

This chapter shows that class inequality either remains the same or, in some respects, has got worse. Class and classism are still strong and continue to have a debilitating influence on individual and community well-being. These conceptualisations of social class and classism allow us to more fully consider the many ways in which working-class people are environmentally disadvantaged, as the following pages testify.

3

Carrying the Environmental Burdens

There is substantial evidence from around the world that working-class, low-income and low-status groups bear a disproportionate burden of environmental toxins and have reduced access to environmental goods, such as green space. This, in turn, impacts health and wellbeing and goes someway to explaining the health inequalities between the working-class and middle-class described in Chap. 2 (for a summary, see, for example, Kruize et al. 2014). The socially unequal distribution of burdens and benefits regarding the environment in which we live, work and play has been called 'environmental injustice' in the US and some other countries, while in the UK it tends to be called 'environmental inequality'. The idea of there being environmental inequalities between different socio-economic groups was accepted and a focus of concern of the most recent UK Labour government from 1997 to 2007. Improvements were made in the monitoring and data gathering related to environmental inequality and, in 2004, the Office of the Deputy Prime Minister introduced environmental factors in its indices of deprivation (ODPM 2004). The Environment Agency, the main environmental regulatory agency in the UK, published several reports on environmental inequality during this period (e.g. Walker et al. 2003a, b). In addition, the Department for

© The Author(s) 2020
K. Bell, *Working-Class Environmentalism*,
https://doi.org/10.1007/978-3-030-29519-6_3

Environment, Food and Rural Affairs (Defra) and other UK government departments commissioned research related to environmental justice, for example, a ground-breaking Friends of the Earth report (Stephens et al. 2001).

However, under Conservative and Conservative/Liberal Democratic Coalition governments, from 2010 to the time of writing, the tide has turned so that most of the relevant research, monitoring and associated interventions have either disappeared or transitioned into forms that are less obviously about tackling inequality. In the only forms of environmental appraisal that have legislative status in the UK—Environmental Impact Assessments, Strategic Environmental Assessments and Sustainability Appraisals—there is now little or no consideration of distributional issues. The Environmental Equality Indicator among Defra's goals for sustainable development was dropped in 2012 (Defra 2013) under the Coalition government. Similarly, the Environment Agency, at the time of writing, rarely mentions 'environmental inequality' in its reports, publications and online pages.

Before looking at the evidence for environmental inequality/environmental injustice on the basis of social class, it is important to note that there is no single definition or assessment measure. We cannot directly measure class as it tends to be a composite of factors, as outlined in Chap. 2. Therefore, we have to look at specific 'indicators', that is, measurable aspects of class, such as levels of education and income. To measure environmental inequality these indicators must be linked to environmental standards to see if there is any correlation, or linked pattern of variation. This is complex because what constitutes a good quality or healthy environment is controversial. Even where standards exist, they often do not consider synergistic and long-term impacts. Different factors are discussed here individually but it is important to remember that there will be cumulative and multiple impacts of poor environments, for example, in terms of the 'toxic cocktails' from industrial sites (see, for example, Bullard 2000; Allen 2003; Lerner 2005). Environmental factors rarely act in isolation but, as the Environment Agency notes, 'cumulative impact assessment is rarely carried out in a formalised or consistent way within EIAs' [Environmental Impact Assessments] (Environment Agency 2007, p. v). In addition, some of the problems and impacts described here may be underestimated due to where and when measurements are taken. For example, perhaps an area appears to have adequate air quality because the measuring devices are placed higher up than the height at which people breathe the air. Environmental standards also usually fail to recognise vulnerable populations. The same level of burdens can have very different impacts on different groups of people, depending on incomes, cultures and background levels of health. Wealthier people can afford to consume better quality food, for example, which may support their body to be more resistant to environmental toxins. Or they may be able to afford water purifiers and air filters or other individual consumer goods that can protect their health from environmental pollutants.

A number of different social classifications have been used to judge environmental inequality in the past, including race and ethnicity, income and gender. Although several studies show that Black, Asian and Minority Ethnic (BAME) people in the UK suffer the worst environmental conditions (e.g. Agyeman 2001; Stephens et al. 2001), the focus of research and policy has tended to be on low income. However, there is a large overlap between BAME and low-income groups in the UK—members of BAME groups often, but not always, live on low incomes. The employment rate for BAME people is 10% lower than the national average and BAME people are almost twice as likely to live in poverty as White people (EHRC 2016).

Taking into account all these provisos around definitions, focus and measurement, this chapter outlines some of the relevant literature, including my own research findings, on environmental inequality on the basis of class. It includes, not only quantitative analysis, but also accounts of working-class people's perceptions of their environments since communities provide a great deal of anecdotal evidence which is not easily captured in quantitative studies.

The evidence, together, shows a clear pattern of inequalities in environments between low-income and wealthier groups, and between working-class and middle-class people in the UK. While I focus here on the UK, evidence suggests that this pattern is reflected around the world, including in Australia, Canada, South Africa, Europe, Latin America, China and South Korea (for references, see Bell 2014 and Kruize et al. 2014). A number of environmental issues will be discussed here, including air quality, accident risk, waste, blighted neighbourhoods, chemical toxicity, food poverty, transport inequities, fuel poverty and flood risk. Each of these will now be addressed in turn.

Air Quality

With regard to air quality, it has consistently been shown that poorer neighbourhoods in the UK tend to experience the worst air quality, and that this directly undermines the health and lifespan of these communities (e.g. Mitchell and Dorling 2003; Walker et al. 2003b; Pye et al. 2006; Milojevic et al. 2017). One recent study, focused on England, concluded that 'Socioeconomically disadvantaged populations often have higher exposures to particulate air pollution, which can be expected to contribute to differentials in life expectancy' (Milojevic et al. 2017, p. 1). There are numerous studies showing the same pattern at a sub-national level, in specific UK cities. For example, Namdeo and Stringer (2008), examining the relationship between air pollution, social deprivation and health in the city of Leeds, found a link between social deprivation and higher NO_2 levels. A more recent study in London found that 85% of the schools most affected by air pollution serve deprived neighbourhoods (FIA 2017). In Wales, this pattern does not occur because deprived people are more likely to live outside the main cities such as in ex-mining communities, where there is less of the human activity that contributes to air pollution (Mitchell et al. 2015).

While much of the poor air quality in the UK is caused by vehicles, there has also been contamination from factories and industrial sites. Hazardous or large industrial sites in the UK are monitored under the Integrated Pollution Prevention and Control (IPPC) system. As well as causing air pollution, IPPC sites can also impact water quality, and create noise, odour and visual disturbance. A number of studies undertaken in the late 1990s and early 2000s showed a strong relationship between the siting of IPPC facilities in England and low-income communities (e.g. Boardman et al. 1999; FoE 2000, 2001; Walker et al. 2003b, 2005a; Wheeler 2004). Boardman et al.'s (1999) study found that more than 90% of the polluting factories in London and 80% of those in the North

East of England were in areas with below-average incomes. Friends of the Earth (FoE 2001) found that 82% of all carcinogenic factory emissions into the air in England were emitted in the most deprived 20% wards. They concluded that 'deprived communities bear the brunt of factory pollution' (FoE 2001, p. 1). Walker et al. (2005a) noted that, as well as major pollution sources being disproportionately located in deprived areas, they also tended to be clustered there so that such areas experience multiple hazards. This pattern has been explained in terms of a tendency to try to protect uncontaminated environments in accord with planning policies and/or the ability of 'articulate', 'well-connected' people, with the time, confidence and skills, to influence planning decisions (SDRN 2004, p. 15). Formal Environmental Impact Assessment procedures tend not to take account of distributional issues. Many of the working-class people I interviewed were aware of air pollution issues in their locality, and sometimes of the inequality in environments and the impact of this, as Anita describes here:

> We have some of the worst air pollution rates in the … country and people in [deprived local area] … are breathing in hazardous dust and particles … We know that it costs the NHS and local government loads of money because we have loads of lung issues in this city that could be avoidable if we were tackling air pollution. (Anita)

Accident Risk

IPPC sites also, in some instances, present an accident risk. A number of the research interviewees were living near to IPPC sites. Because of the potential risk of accident involved in living near these sites, in one area, the local Council issues residents with a leaflet regarding the industrial related hazards in their area called 'What should you do if you hear the … Sirens?' It explains: 'The … Sirens warn people of possible incidents involving dangerous substances … The … Siren System is maintained by an independent charitable organisation … The sirens are tested monthly on the third day of each month at 3pm' (p. 1). If people hear the sirens outside of this time they are told to stay indoors and not leave the area. The companies mentioned on the leaflet as producing hazardous substances include a pharmaceutical company (described as producing 'flammable and toxic substances'); a petroleum company ('extremely flammable and dangerous to the environment') and a liquefied natural gas company ('extremely flammable'). Inside the back cover, on page 10, there is an invitation to get a translation of the leaflet in other languages. It is unlikely people would get that far into the leaflet, so possibly speakers of other languages would miss this vital information. Ange, who drew my attention to this leaflet, told me:

> *Every three months they do a bell alarm for the area and we know it's on at 3 o'clock in the afternoon … If it happen any other time … they give us a leaflet to say you stay home, don't go out and if you have the children at school don't go pick them up, you know, the staff there's going to look after the children … And one time I heard the bell, the siren and it wasn't at 3pm, so I ask* [the local councillor] *and her answer was: 'oh it's nothing, no big problem, you can go on the website and check yourself'.* (Ange)

The idea that the emergency service is being run by a charity rather than as a statutory service, that speakers of other languages or illiterate people may not get this information, that people without internet would not be able to check the severity of the risk and that the emergency siren goes off for 'nothing' seems to be an inadequate approach to people's safety.

There are fewer industrial sites in the UK now as much of our manufacturing (and, therefore, IPPC pollution and accident risk) has been outsourced, that is, shifted overseas. This pollution and risk will now impact the poorest communities internationally. According to the latest data from the World Health Organization (WHO 2018b), pollution inequality between the world's rich and poor is widening with seven million people—mostly in low-income countries—dying every year from airborne contaminants.

Waste Management Contamination

IPPC sites also generally include landfill and waste management sites. These sites can cause health and safety problems as a result of traffic, noise, odour, vermin, air pollution and toxicity of local land and water. A number of studies indicate that those living near landfill or waste treatment sites are at increased risk of a range of health problems including low birth weight, cancers and respiratory illnesses (e.g. Vrijheid 2000; Saunders 2007). There appears to be class inequities with respect to the siting of landfill sites in a range of countries. For example, a recent review of the literature on waste management covering Europe and the US found 'consistent indications' that waste facilities are disproportionally sited in more deprived areas and that 'This applies to waste incinerators, landfills, hazardous waste sites, legal and illegal' (Martuzzi et al. 2010, p. 21). In the UK, though there is mixed evidence, it has also generally been found that those living closest to landfill are more likely to be socio-economically deprived than those living further away (e.g. Wheeler 2004; Fairburn et al. 2009; Richardson et al. 2010). Research indicates that the reason for this is sometimes that lower income areas are disproportionately selected for siting of these facilities but also that wealthier people tend to move away from landfills if sited near them (Richardson et al. 2010).

Now that landfill is being used less, we are finding environmental equality issues associated with recycling. The working-class people I spoke to during my research frequently mentioned recycling as having a negative impact on their lives, from having to deal with recycling at the household level (see Chap. 4) to the toxic recycling facilities that they have to live with. Those living near such facilities said that their health and safety were not adequately considered in the siting or management of these facilities.

For example, in Bristol, where 31% of residents have a limiting long-term illness, health problem or disability (BCC 2017a), compared to just 16% in the wealthier areas (BCC 2017b), residents spoke about a number of local environmental harms related to recycling. Some of these issues have also been reported in the local and national press. For example, the residents told me, and it was also reported in the media (BBC 2017), that a recycling company had been found guilty of dumping 61,000 tonnes of toxic waste into the foundations of a supermarket distribution centre in the area. The recycling company workers reported feeling dizzy and sick when moving the material, but their complaints were ignored by the managers. Eventually one of them alerted the Environment Agency which brought about an investigation and eventual prosecution. The crime included falsifying documents and knowingly consenting to illegally dumping controlled waste without a permit. The recycling company has since changed its name.

In the same area, there were incidents resulting from bales of waste awaiting export being stored near homes. On several occasions in 2011 and 2014, the bale wrapping became torn, allowing fluid to leak out which caused severe fly infestations. The company involved, which produced refuse-derived fuel (RDF) from household waste, was eventually fined £18,000 by the Environment Agency and was shut down for several months. However, it was later given a new contract (for wood recycling) (BBC 2015). In another incident in the area, a company which recycled steel and plastic drums breached its environmental licence on several occasions (Environment Agency 2015a).

Despite all these problems, in 2015, the Environment Agency claimed that air quality in the area is typical of an urban setting and should not give rise to increased risk of respiratory health problems and that the companies were 'currently compliant with their environmental permits' (Environment Agency 2015b, p. 1).

The interview participants from this area and around the country highlighted the difficulties associated with waste management facilities being located in their local area. For example, Paul was very concerned that these facilities were having an impact on his family's health as well as that of the local community. He said:

> When the chemical companies have finished with the drums they send them to …. [recycling company] which is 55 yards from the children's bedroom and they've got an incinerator stack there which they burn out the residues … I've grown up around the … area and a lot of people my age are dead and a lot of people my father's age, my father died at 42 from a massive heart attack and stroke and lots of people are dying from heart attacks and strokes. Lots of people are getting brain tumours, lots of people are getting kidney disease … and you know, we've got high instances of birth defects, we've got major issues with asthma, eczema etc., from a lot of the stacks that are … are belching out. (Paul)

These recycling facilities are also, sometimes, noisy. Noise is associated with increased blood pressure and heart rate, and higher levels of stress

(Ising and Kruppa 2004). Paul complained about noise which he believed emanated from a metal recycling company with a base in the area where he lives. A local community group Facebook page contains numerous posts from residents who are disturbed and stressed by the frequent sound of metal being broken up, including occasional explosions. Paul described the impact of this:

sometimes you haven't had any sleep for four or five days because [the noise is] *above World Health Organisation levels and you're trying to not kill your children, make sure they get to school, make sure they come back from school, feed them, do your work in between ... It's quite hard to do and it does stress you out. There are times where you are completely stressed.* (Paul)

So residents in these deprived areas are experiencing a whole range of problems associated with recycling, some of which they say is impacting on their physical and mental health. Although some of the concerns raised by residents and workers locally have led to prosecutions and/or publicity, when we consider the poor health in the area and the difficulty of obtaining information and proving toxicity, these may just be the tip of the iceberg.

Blighted Neighbourhoods

Alongside these major disturbances to health are a range of more insidious environmental problems that add to the blight of working-class areas and more subtly undermine health and wellbeing. Problems such as litter, graffiti, dumped cars, fly-tipping, dogs mess, vandalism, derelict land, broken glass, potholes and broken pavements are more likely to be found in working-class area. Though apparently minor, these problems can have impacts ranging from an undermining of wellbeing due to unpleasant views, noise and smells through to more serious health issues. Research indicates a link between the presence of these blights and poor health outcomes (Ellaway et al. 2009), with people who perceived high levels of these problems in an area being more than twice as likely to report frequent episodes of anxiety and depression than those who perceived low levels. This phenomenon was stronger among social groups who spent more time in the local area, such as older people (Bowling et al. 2006; Mottus et al. 2012). In addition, fear of crime can increase when the local environment does not appear to be cared for, leading to more isolation as people fear leaving the house (SDRN 2004, p. 26). Furthermore, vermin and associated disease may be attracted to litter and rubbish. These lower level environmental issues are often considered to be the fault of the local residents. However, as Chap. 4 discusses, such problems can often result from ineffective policy making and inequity in environmental services. For example, in more deprived areas, people generally have a lower standard of street cleaning and waste collection services (MacIntyre et al. 2008). Some of the people I spoke to in my research perceived an inequality of environmental services across the city, with wealthier areas seeming to be better cared for. For example, Jim, from a predominantly working-class and multi-ethnic inner-city area, said of his neighbourhood:

> *Environmentally, I think it could be a lot better. It's next to the … [motorway] … so it's not a good environment … Then, the other thing is, … [the local area] … doesn't get the manicuring that, say, … [the wealthier area] … or those other areas get, so yeah, it always looks a bit down on its heels.* (Jim)

The blight found in working-class areas and associated harm is exacerbated by a lack of green space. Green spaces provide numerous benefits in terms of the possibilities for recreation; aesthetics; impacts on physical and mental health; developing community; noise regulation and air pollution reduction. Engagement with the natural environment, in general, is believed to foster healthy neighbourhoods, encourage social activity, reduce crime, help children's learning and involve people in sustainability initiatives (Marmot 2010). In 1984, a study showed that hospital patients recovering from operations are likely to recover quicker and need less powerful painkillers if their hospital bed allows them to look out onto a natural scene (Ulrich 1984). Since then, a whole body of evidence demonstrates the importance of contact with nature for physical and mental health and wellbeing (e.g. Mitchell and Popham 2008; Maas et al. 2009; Marmot 2010). There are particularly significant impacts on children's, older people's and low-income group's physical and mental health (Sustainable Development Commission 2010; Maas et al. 2006; RSPB 2010). Access to attractive local walking areas, such as public green space and footpaths, is likely to increase the amount of walking that the people in the neighbourhood do (Maas et al. 2009). Green spaces also encourage more socialising between neighbours, increasing caring and supportive behaviour (Defra 2011). The closer that green space is to where we live, the more likely we are to use it (Bateman et al. 2006), and the noise abatement and pollution reduction aspects of green space also tend to be enhanced with greater proximity (Brown et al. 2011, p. 390). Therefore, local access to green space should not be considered a luxury but an important component of proactive health policy for those living in urban environments. It is not enough to have a few lovely parks in our cities and towns; we need to see green space when we go about our everyday activities to most benefit from it.

A number of studies and reports show that low-income groups and/or people living in deprived areas (as well as people from minority ethnic communities, older people and disabled people) have less access to green space (e.g. Defra 2011). Working-class people use green space much less than other sections of the population (Burt et al. 2013). Those who most visit green spaces are middle-aged, employed and from ABC1 socio-economic groups (Burt et al. 2013). Compared to the UK adult population in general, people living in deprived urban areas were 40% less likely to visit natural areas and green space than those living in wealthier urban areas (Defra 2011). There may be many reasons for this, including lack of time, quality of local spaces, fear of crime and affordability in terms of getting there and, increasingly, entrance fees. Almost half of the country's most socially deprived areas are more than 15 miles by road from 10 national parks and 46 Areas of Outstanding Natural Beauty (CPRE 2018). The Campaign to Protect Rural England has called for better bus and train links from towns and cities to improve access for disadvantaged groups.

Some of my research respondents spoke about these issues. Here, Sharon and Mel describe the inadequate upkeep of their local green spaces and an increasing tendency to have to pay to use them, sometimes as a result of austerity measures.

I would really like to use those [park sports facilities] *but they haven't been repaired ... there's quite a lot of smashed bottles or glass around, and the same for the children's play area. There's a couple of things like a kids' mini trampoline but you can see that there's glass and bottles underneath there, so you're kind of like anxious for any child to use it. I don't like taking my nephew there in case of what might be around in the play area.* (Mel)

In [local green space] *there isn't really a safe place for the young kids to play football ... since the cuts* [to services and budgets] *they're making more meadows now, so you don't have to cut the grass. So the little football area where the kids used to play off the estate has gone because it's now meadow land. They can be charged to go and play in, like a football area ... There's more people charging because the grants that were available are now going.* (Sharon)

Green spaces, because of their benefit and necessity for humans, are often described as 'environmental goods'. As such, steps need to be taken to overcome these barriers to working-class access. Other environmental goods include food, energy for heating and cooking, and transport. In all these matters we, again, see environmental inequities with working-class people being relatively deprived and sometimes forced to go without, as the following sections describe.

Food Poverty

The main reason working-class people tend to lack environmental goods is because they are more likely to live in poverty or situations of economic insecurity. While not all working-class people are poor, most poor people are working-class, so poverty is mostly a class issue. Due to unemployment, underemployment, casualisation of work, low wages, benefit cuts and community breakdown, poverty now impacts many working-class people. The Joseph Rowntree Foundation (2018) recently published a report on living standards and household budgets in the UK, finding that families living on low incomes are significantly worse-off than they were ten years ago. The study on which it was based found that full-time earnings on the minimum wage are not sufficient for a basic standard of living

and increased prices for housing and transport are compounding deprivation. In the city where I live, Bristol, often topping the polls as one of the
best places to live in the UK, more than one in four children are living in
poverty (27.8%) (BCC 2019). Like many countries in the world, UK
poverty is health impacting and even life threatening. This is particularly
the case with regard to food poverty and its consequences.

Food poverty or food insecurity is a growing UK problem. The United
Nations estimates 8.4 million people in the UK now experience food
insecurity. This means having 'limited or uncertain availability of nutri-
tionally adequate and safe foods or limited or uncertain ability to acquire
acceptable foods in socially acceptable ways' (USDA in Taylor and
Loopstra 2016, p. 3).

The related undernourishment mainly increases illness and death rates
in a slow and hidden way, although there have been recent reports of
individual people starving to death (e.g. Gentleman 2014). Almost half
of workers are now worried about being able to afford to eat and one in
eight workers struggle to afford food, skipping meals to make ends meet
(TUC 2017). In Bristol, in 2016/17 more than 5452 children depended
on food charity just to eat, an increase from 4805 the year before. In
some areas of Bristol, including Lockleaze where I live, half the children
struggle to get three meals a day (End Child Poverty 2016).

It has been well documented that parents in low-income households
go without food so that their children can eat (e.g. Davis et al. 2016). The
demand for food banks continues to rise, year on year. There are at least
2000 food banks now running in the UK, distributing emergency food
to those who cannot afford to buy it. The Trussel Trust, the UK's biggest
food bank network, reported that it gave out a record 1.2 million food
parcels in 2016–2017. Some people have argued that people go to food
banks for freebies that they could afford to buy if they had to. Those who
make these claims do not see the use of food banks as evidence of hunger.
However, most food banks require their users to obtain a voucher from
an outside agency which validates their need for the food aid. The overwhelming evidence is that people only generally use them when they are
desperate (see Garthwaite 2016). Lisa Mckenzie (2015) reports that the
women in her research on the council estate where she lives said they

would rather 'go on the game' (i.e. become a prostitute) than go to a food bank.

Food charity mainly deals with the most desperate, but there are many other working-class people who have enough food, but not of good quality. Many, and not just those on low incomes, would struggle to pay for a healthy or organic food diet, may not be accustomed to it or may not have it readily available. For example, the women I spoke to in my research said:

> *Fruit, I find difficult. I only had fruit once a year when I was growing up and that was at Christmas. I didn't have fruit in between, apart from tins of fruit cocktail. It was expensive … so we didn't have it. Didn't have fruit so it's not part of my diet now…I'd buy fruit for my daughter but not necessarily for me.* (Sharon)

> *The area where we live we have a Co-op but … we don't have much of, you know, fresh food. Everything in there … it just makes people fat and then after you eat you're hungry again straight away because it doesn't have any nutrients, you only got calories … If you look around in the area, there's a lot of overweight people… they eat lots of carbs because bread is cheap, fills you.* (Ange)

> *It's all packet food now. You could buy fresh veg and what have you, you can cook it and it's probably cheaper. But a lot of mums are hard pressed and people are just trying to do, say, a job here and a job there and trying to fit in different hours and they haven't got the time. I know that seems a bit of a pathetic excuse 'cos sometimes it don't take that long to cook a dinner, but also it is the time.* (Irene)

These statements also provide a clue as to the reason behind the so-called 'obesity epidemic', thought to be killing nearly three million people a year worldwide. While the policy response tends to be condemning the victims for their poor choices, the issue clearly links to poverty and associated stress (although poverty can also result in being underweight) (Hughes and Kumari 2017). Inadequate diets could also explain the poor health outcomes between different wards in Bristol. For example, in Hartcliffe and Withywood in Bristol, 42% of residents are living with a limiting long-term illness, health problem or disability

(BCC 2017c), compared to just 16% in the wealthier ward of Clifton (BCC 2017b).

Chemical Toxicity

Chemicals in pesticides, coatings, printing inks, adhesives, cleaning agents and personal care products continue to increase in range and volume. In a recent paper, McDonald et al. (2018) showed that the volatile chemical products (VOCs) found in these products now contribute one-half of emitted VOCs in 33 industrialised cities. According to the Lancet Commission report discussed in Chap. 1, 'chemical pollution is a great and growing global problem. The effects of chemical pollution on human health are poorly defined and its contribution to the global burden of disease is almost certainly underestimated' (Lancet Commission 2017, p. 1). The report informs us that more than 140,000 new chemicals and pesticides have been created since 1950. Fewer than half of the most regularly used chemicals have undergone any testing for safety or toxicity and these have 'repeatedly been responsible for episodes of disease, death, and environmental degradation' (Lancet Commission 2017, p. 1). Some of the newer chemicals include developmental neurotoxicants, endocrine disruptors, herbicides, insecticides, pharmaceutical wastes and nanomaterials (Lancet Commission 2017, pp. 1–2). Even when there is a testing and licencing regime in place, companies still regularly make and sell products that are potentially dangerous to humans (see Chap. 7 on glyphosate).

Exposure to these chemicals may be causing a silent pandemic of disease and impairment. For example, research undertaken by the Alliance for Cancer Prevention and the UK Working Group on the Primary Prevention of Breast Cancer (2013) indicates that lifelong, low-level everyday exposure to toxic substances and hormone-disruptors—from pesticide residues in food to chemicals in cleaning and body products and in the workplace—is linked to the ever-rising rates of breast cancer.

As a result of lower or more insecure incomes and fewer choices, working-class people are also more likely to be exposed to toxic chemicals, either in the atmosphere through manufacturing such products in toxic factories, or through their everyday use in relatively cheap house-hold cleaning and body products. For example, working-class people may be generally less able to afford organic alternatives or to access information about the harm caused by the chemicals in everyday products. Most of the working-class people I interviewed for my research were unaware of these issues, but some had learnt about them because they had been ill and decided to explore the topic. However, there was little trust in the information that was available. For example, Dave and Jo said:

I've started to realise recently that, you know, these things [the chemicals used in everyday life] *start to have an effect on you, you know, and I try and diminish the amount I use as far as possible, because I think we need a proper examination of a lot of this stuff, because ... I think these things could be having far more of a profound effect than we actually realise. But the question is, who's going to do the investigation? Who's going to be truly independent to actually call these people out, if there is a problem?* (Dave)

I don't know much about how the environment affects our health...I know when I start using bleach, spray or whatever, that I am coughing like a mad man and it is probably not good ... Then I found out you can clean your house with white vinegar, apple cider vinegar, toothpaste, tomato sauce on silver. It is amazing. All these things, no chemicals and you get a better result. (Jo)

The issue of chemical impacts on the health of working-class people is under-researched, though some information has been made available through Trade Union Health and Safety investigations (see Chap. 6).

Fuel Poverty

When people struggle financially, they may have to make choices about whether to eat or heat. Many working-class people, especially older people, can't afford to put their heating on when they need to, increasing their risk of colds, flu, chest infections, respiratory problems and hypothermia. Sometimes this can be fatal. Many people die in the UK every winter because of cold. The average deaths in winter compared to preceding and subsequent months are called Excess Winter Deaths. In 2014–2015, there were an estimated 43,900 Excess Winter Deaths in England and Wales and 24,300 in 2015–2016 (ONS 2017). Many of these will be a result of energy-saving measures taken by people who cannot afford to heat their home. It is not just because the UK is a cold country as we experience a higher number of excess deaths than in colder countries such as Norway and Sweden.

In the UK, 'fuel poverty' is now defined as a situation where, after spending money on adequate heating and other energy requirements, a household's income would fall below the official poverty line. According to the UK government's own statistics, more than 10% of households in England are living in 'fuel poverty'; this includes 20% of those living in private rented property and 25% of single parents with dependent children (DECC 2016). Levels of fuel poverty are even higher in other parts of the UK, at 42% in Northern Ireland, 35% in Scotland and 30% in Wales (National Energy Foundation 2017). Rates of fuel poverty also differ across small geographical areas. In Bristol, for example, approximately 13% of all households live in fuel poverty, but in the worst affected ward, 27% of households were considered as being fuel poor (BCC 2015, p. 9). Some food banks also operate fuel banks, which provide vouchers for pre-payment meters. Surveys show, for many recipients, the vouchers are an option of last resort: around half had self-disconnected from their gas and/or electricity supply when they received the voucher. Most (80%) of fuel bank users are on very low incomes of £10,000 or less (NEA 2017). The government targets for reducing fuel poverty may not be met until 2091 at the earliest, according to a recent report (IPPR 2018a).

Even if not in fuel poverty, many working-class people may be experiencing fuel deprivation that impacts their health and wellbeing. A recent Trades Union Congress (TUC) report (2017) found that one in six workers had left the heating off when it was cold to save on energy bills. Almost half were worried about being able to pay their energy bills.

Fuel poverty and fuel deprivation occur, not only because of low incomes and poor energy efficiency but also high fuel prices. Energy companies make large profits, with the UK's six biggest energy companies making a profit of £1 billion in 2016 (Ofgem 2017a). The Conservative government's answer is to encourage consumers to switch companies. Yet, working-class people are much less likely to switch supplier than middle-class people, in part because the transition can tie up sums of money, as discussed in Chap. 4. Approximately 17% of middle-class people (social class AB) switched energy supplier in 2011, compared to just 8% of Social Class E (Ofgem 2012). The topic of energy justice and fuel poverty is further discussed in Chap. 4 under environmental policies.

Transport Inequities

Inadequate transport access has been called transport poverty, transport exclusion and transport inequity. Transport is an equity issue in part because cheaper, more affordable housing tends to be located in areas with poor transport connectivity as well as because transport can be too expensive for people living on low incomes. Lucas (2012) identifies the main aspects of transport disadvantage as limited access to travel (limited access to provision of public transport or a car), prohibitive fare costs, and limited, or non-existent, travel-related information. These have social and health impacts because having efficient and affordable transport enables people to access essential and valued opportunities such as jobs, education, shops, healthcare and social networks, while a lack of transport mobility is linked to social disadvantage (Lucas 2012) and isolation (Currie and Delbosc 2010). Transport poverty prevents people from participating in the economic, political and social life of the community (Kenyon 2006).

The most appropriate public transport is essential for working-class people's mobility and health. Working-class people tend to walk and use the bus while wealthier groups tend to use rail and cycle (Titheridge et al. 2014). Therefore, as a principal means of mobility for working-class people, bus services need to be reliable, punctual, affordable, safe, non-polluting, accessible and, ideally, comfortable. In particular, they need to meet the needs of those working-class people who most depend on them—children, young people, older people and disabled people. However, in many parts of the UK and the world, bus services are sadly lacking. One of the people I interviewed, for example, explained the difficulties she faced in trying to access the local bus service from where she lives, in a deprived part of her city:

> *Because I don't drive, I've got to travel a lot by bus but the transport in … [local area] … is really poor … so sometimes I end up walking but then I've also got back problems, so I'm putting myself in pain by walking sometimes, so it's kind of either take the time or take the pain. Last week I came into work. It took me two and a quarter hours to get from* [her local area], *so if you think about the impact that that has on your time … It's ridiculous! … They must have spent a fortune doing all these new digital bus stop signs. But they lie … I've walked past ones before that said, you know, it's another 40 minutes to the next bus. You start moving away from the bus stop and then it arrives, so you then miss that.* (Mel)

The interviewees also spoke about the difficulties in accessing rural areas and transport issues for rural inhabitants, as Anita described;

> *… people are obese and they need to go out walking more but local transport's died.… There are loads of spots in* [nearby countryside] *…. that I would go walking every weekend but I can't get there now because there's no transport. There's no buses. There's no public transport. How that impacts on the rural working-classes, I can't even begin to imagine when you can't get to your job centre appointments, you get sanctioned for your Universal Credit or you can't get to a supermarket and your local shop's probably going to be more expensive, all those things, but it also means people can't get into the countryside.* (Anita)

This anecdotal evidence illustrates the impact of a national trend where, in England and Wales, Council bus budgets have been cut by 45% since 2010 and subsidies for buses have fallen by £20.5 million in the past year (Campaign for Better Transport 2018). At the same time, household transport costs in the UK have risen from an average 10% of household budgets to almost 20% (Joseph Rowntree Foundation 2018).

Because of poor bus services and a greater likelihood of living in cheaper housing on the outskirts of urban areas, far from work, leisure and service opportunities, some working-class people can be very car dependent. However, generally, it is the middle-classes that drive more. The UK National Travel Survey (NTS) (DfT 2013) showed that, in

2012, wealthier people were more likely to own one or more cars, make more car journeys and take longer car trips than those with lower incomes. A further equity issue that arises here is that, while working-class people are less likely to drive, they are more likely to be killed and injured by cars.

In 2016, there were 1792 road deaths, and 181,384 road casualties in the UK (DfT 2017), with working-class children being much more likely to be killed or injured in road accidents than their middle-class counterparts. For child pedestrians, rates of serious injury are four times higher in the most deprived areas than in the least deprived (Edwards et al. 2007; Siegler et al. 2010). The main factors that put people in disadvantaged areas at risk of being involved in road traffic accidents include living in more hazardous environments, such as near high volumes of traffic; lacking a car, so being more likely to walk; and a lack of safe places for children to socialise and play, so they are more likely to use the roads (Lowe et al. 2011). Outside of the UK, there is international evidence of a similarly strong relationship between deprivation and road casualties (e.g. Braver 2003; Hasselberg et al. 2005; Factor et al. 2010).

Flooding and Disaster

As well as being in greater danger from local pollution and traffic, working-class people are also more likely to be impacted by all the risks associated with global climate change, including flooding and other disasters. In class societies, the wealthier are less likely to be impacted by climate crisis because they can change their dwelling places, jobs and other aspects of their lives in the face of risks and catastrophes. In the UK, this inequitable burden is evident in a key risk that is exacerbated by climate change—flooding. The risk of flooding is more evident in proximity to rivers, coastlines and other water bodies and, in the UK, it is those on lower incomes that are more likely to be living in these areas, particularly coastal areas where housing may be cheaper (Walker et al. 2005b, 2007; Walker and Burningham 2011). Research by Oxfam showed that poorer areas have been at least three times more likely to be flooded than wealthier areas in the UK (Oxfam 2014). As well as being more at risk, working-class people are also more vulnerable due to less access to information, poorer quality housing, less flood protection and less ability to recover losses (Fielding and Burningham 2005; Whittle et al. 2010). In particular, they are also less likely to have contents insurance since, after spending on other household costs, there may be nothing left for non-immediate expenses such as insurance premiums. Flooding impacts are also compounded by intersectional issues such as age and disability.

In addition, working-class people may experience less help and support when flooding does occur. Hurricane Katrina was a classic example of inadequate and inequitable preparation and response to flooding, based on racism and classism. In the UK, Johnson et al. (2007) analysed flood risk management policy finding that they 'fall far short of being fair from either a vulnerability or equality perspective' (2007, p. 1). The researchers found that economic efficiency considerations were strongly prioritised over equality principles. One of the interviewees for my research, Dave, made the following point about the economic drives behind flooding, a seemingly 'natural disaster':

I do feel deeply sorry for the people that are flooded, and I feel particularly sorry for the people that have been, in my view, mercilessly conned into buying houses on flood plains, where the people who sold them those clearly knew without any excuse what they were doing when they set those people up. They're caught in a trap because they'll never sell those properties. No matter what they do they're just waiting for the next floods to come. (Dave)

The health impacts of flooding range from immediate risk of injury or death; illness resulting from the associated hygiene issues and living in damp accommodation (e.g. gastroenteritis, respiratory disorders); to psychological distress (e.g. anxiety, panic attacks, agoraphobia, chronic anger and depression). Of course, middle-class people will also suffer terribly as a result of floods and may have intersectional issues that make them particularly vulnerable, but the burden of flooding will generally fall on working-class people. Climate change means that flooding will become more severe and/or more likely. Sea level rise and more extreme storm events will impact the coastal areas and, since there is a disproportionate concentration of deprived populations living in those areas in the UK and some other countries, this will impact more on working-class people.

On a global scale, though, it is important to remember that those most vulnerable to climate change are people living in low-income countries where there is limited infrastructure to protect and support them (UNHDR 2015).

Double Injustice

The above discussion illustrates how working-class people carry the environmental burdens of society. The double injustice is that working-class people are also the least likely to create environmental harms. We know that there are differing consumption patterns between working-class and middle-class people, with greater consumption being associated with higher carbon emissions, pollution and more use of irreplaceable natural resources. The greatest lifestyle contributors to climate change are driving cars, flying, eating meat and having children in the high-consumption developed world (Wynes and Nicholas 2017). Globally, it is the wealthier that are most likely to engage in these.

Whilst climate change is more likely to occur as the result of actions undertaken by the middle-class and affluent, the impacts of those changes will fall on the working-class and the least well-off. The wealthiest 10% of people in the world are responsible for around half of the CO_2 emissions attributed to individual consumption, yet poorer people are the most vulnerable to the risks associated with climate change (Oxfam 2015).

This chapter gathers together evidence that strongly suggests that it is working-class people, globally and locally, that are the most likely to experience poor quality environments. Sometimes working-class people lack the means necessary to avoid environmental harms or access environmental benefits or they may even be the objects of outright environmental classism. The following chapter discusses some of the reasons for these inequities, including policies which do not consider the different needs of working-class people or even ignore their existence.

4

The Environmental Policy Makers

The recent *gilets jaunes* or 'yellow vest' protests in France were sparked by a planned rise in 'green taxes' on petrol and diesel. Though frustration about other issues also contributed to the protests, they have drawn attention to the potential for climate change mitigation policies to impact negatively on working-class people and, thereby, potentially alienate them from the cause overall. Without a social justice perspective, environmental policies risk exacerbating the gap between rich and poor, and climate mitigation and other sustainability outcomes may be hard to achieve.

As this chapter outlines, environmental policies, services, improvement programmes and transition processes—locally, nationally and internationally—have often forgotten about or excluded working-class people and low-income groups, sometimes compounding their disadvantage. Environmental policies sometimes seem to be developed without taking into account their social impacts, or sufficiently considering the capacities necessary for members of the public to implement, comply with or engage with them. This inevitably discriminates against working-class people who are more likely to feel burdened by policies that negatively impact on those with lower incomes, less free time, poorer physical health and greater levels of stress.

© The Author(s) 2020
K. Bell, *Working-Class Environmentalism*,
https://doi.org/10.1007/978-3-030-29519-6_4

Therefore, environmental policies should be evaluated, not only for their efficacy in alleviating an environmental problem in a particular place, but also for their impact on inequality and whether they will reduce or worsen the inequities already faced by working-class people. Research suggests that peoples' willingness to make changes so as to curb their own carbon emissions is significantly affected by their sense of how fairly the burden of action is shared across the population (CSE 2018). Achieving 'fairness' or 'justice' in climate and other environmental policy may, therefore, be key to inspiring public action on climate change and other environmental issues.

Although there are vast amounts of environmental policies to consider, the discussion here will focus on seven key policy areas: planning and urban regeneration, street cleanliness, waste collection, pollution policy, energy policy, green jobs and green programmes. As with Chap. 3, the discussion focuses on specific problem areas of policy. In Chaps. 7 and 8 we will turn to the many potential and existing successful policies.

Planning and Urban Regeneration

Planning and urban regeneration policies are often touted as improving the lives, health and prosperity of local residents. There is some research evidence suggesting that improvements do sometimes occur (e.g. Batty et al. 2010; Thomson et al. 2013; Egan et al. 2013). In the UK, there is a seemingly fair and democratic system of environmental decision-making in relation to planning and urban regeneration. The Town and Country Planning system regulates local land use, including new building works, through Development Plans which involves consulting the public. Any new developments proposed after the Plan has been agreed are subject to planning permission which also has to take account of public opinion.

However, there is also evidence that the outcomes and processes involved in planning and urban regeneration are not as fair and beneficial as might seem. Planning consultations can be restricted prior to the public discussion and this can disadvantage working-class people. For example, a study of strategic land use planning in a county in England found that, before the public consultation process even began, local political representatives had decided that the wealthier areas were to be protected from large-scale redevelopment (Abram et al. 1996). Some aspects of regeneration, such as the construction of new main roads, can have a negative impact on deprived and working-class communities, increasing noise disturbance and creating severance from facilities or social networks (Hart 2008; Foley et al. 2017).

Some 'regeneration' also leads to what Ruth Glass (1964) has called 'gentrification' with working-class people displaced by professionals in spaces they had once considered their own. Gentrification commonly occurs in urban areas where prior disinvestment in housing or locality creates opportunities for profitable redevelopment. The needs of middle-class incomers or visitors are often met at the expense of the local working-class with social housing and amenities that working-class people use tending not to be a priority in these scenarios. Jim perceived this to be the situation with regard to a shopping centre development near his home in a deprived inner-city area:

> *They had a mock-up of what … [the development] … would look like, and when we asked 'Where's the social housing?', because one thing a local authority can do, they can bend the developer to meet some of their own needs, and they said, 'Oh, yes, it's over here,' and when we looked at the model, what it was, it was just attached to the [multi-storey] car-park. So the back and the sides had no view at all, just the car-park… I mean, it just seemed terrible that you could get to the twenty-first century and build social housing as part of a car-park. I was really kind of surprised at the lack of concern that the developers and Council had for the local people. (Jim)*

There can also be less dramatic and rather more gradual gentrifying of areas where middle-class people buy up the houses that working-class people vacate. As such areas change we can see bistros, coffee shops, delicatessens, organic grocers, health-food shops, bicycle shops and expensive boutiques appearing, gradually changing the feel of the area. The topic of gentrification is well covered in the literature (e.g. see Slater 2011 for an overview) and some might argue that this kind of change could be an improvement to an area, introducing some healthier options and economic vitality.

Certainly, if we consider what is happening in some of the working-class areas that are not yet being gentrified, this might be considered to be a preferable alternative. In some working-class areas there appears to be deliberate targeting by polluting facilities or commercial outlets that risk damaging health and wellbeing. For example, the shops in working-class areas are often full of fast food takeaways; pawnshops; betting shops; tanning salons; cheap alcohol and tobacco shops; and money lenders offering instant cash at high interest rates (Townshend 2017). Research indicates that the food environment (i.e. the food we see being sold and consumed around us) influences our diet (e.g. Caspi et al. 2012). It is not surprising, then, that working-class people are more likely to consume processed, high-fat and sugar-rich foods all linked to obesity, type-2 diabetes, coronary heart disease and cancer. There is a direct link between neighbourhoods with higher concentrations of unhealthy food outlets and obesity in older children (Cetateanu and Jones 2014). Amy, living in one of the most deprived towns in the UK, commented on the changes in the local area and the consequent difficultly of obtaining healthy food:

> There is a lot of fast food … KFC, McDonalds … Like most places we have out of town shopping centres and superstores and things like that…but, in town, we used to have quite a good market on the triangle but that's disappeared … Things like fishmongers have finished, butchers have gone, our little family butchers have had it now … they can't compete with the places like Tesco and the out of town stores like Asda. So yeah, it's not brilliant. We've got a few little bakers, independent bakers still going. Not everyone wants to eat cake and pastry though, do they? (Amy)

The chains of booze shops in working-class areas may have contributed to the demise and loss of the local pubs. These have declined hugely in numbers in recent years, part of a general decline, with more than 25% of all pubs lost between 1980 and 2016 (BBPA 2017). The booze shops are harmful because people can drink more when they are not being observed and the alcohol is not being measured. One of the research participants, living on an estate where the last remaining pub to serve the estate of 5000 houses was going to shut within the next few months, lamented:

There's nowhere for people to go now the pubs are shutting. A lot of them near me [neighbours] *are on their own since they lost people, so they like to go there where they can see other people they've known for years. And they save on the heating when they're down there. So soon they'll have nowhere out of the house to go.* (Joan)

To the extent that planning permission is given for so many of these unhealthy outlets in deprived areas, local planning processes sometimes appear to favour business interests at the cost of working-class communities. In my experience, the views of working-class people can be dismissed in development consultations when their message is counter to the vested interests of the business sector and local or national authorities. For example, as a member of an environmental campaigning group, made up predominantly of working-class people, I took part in a three-year local government consultation process on a Neighbourhood Plan for our area which would shape the wider Local Plan and influence a proposal to build an undisclosed number up to 7000 houses on the estate (more than doubling the density of housing on the estate). House prices were soaring at that time and the local Council wished to sell off the land to private developers. Our group opposed the scale of the development on the grounds that it was not necessary and that it could harm local people, including those who bought the houses since some would be sited near electricity transmission lines. Homelessness in the UK often occurs because people can no longer afford to pay the rent or mortgage on the home they'd had, not because there is no building available (Shelter 2013). In addition, there were, in that year, 4519 empty homes in the city (in 2017 this had still dropped only to 4422) (UK Gov 2018). Taking into account that many on Local Authority waiting lists are currently renting from buy-to-let landlords, there were probably more than enough houses already in existence in the city, if housing were organised according to need. We considered the development to be more about profit than need and also objected, on health grounds, to any houses being built near the high-voltage transmission pylons that run through the estate.

During this period, I witnessed my own and other working-class people's opinions on planning and development issues frequently being dismissed by local authority officers. At weekly meetings over the three-year period, we continually raised a number of issues, including the need to preserve the small green spaces near our houses for children's play; and the dangers of building near the pylons. The electro-magnetic frequencies produced by these facilities have been linked to cancer, miscarriages and depression (e.g. childhood brain tumours, Kheifets et al. 2010). Despite us bringing research-based and local anecdotal evidence to the discussion, our input was consistently dismissed by the local Council and the minutes of the meetings often failed to record our comments. At the end of the three-year period, the local authority officer told us that none of our input would be integrated into the planning document and that we could only supply a written response which would be included as an appendix item (BCC 2011).

Hence, our contribution was marginalised and, in effect, disregarded. Having attended these meetings, sometimes weekly, for the entire three years, the local activists, almost all working-class people, were consequently left feeling very demoralised, frustrated, hurt and angry. Many of our members refused to participate in further environmental decision-making processes; our group lost its momentum; and the consultation process caused divisions in the community because no attempt had been made to integrate the different views. This story illustrates that an opportunity to participate does not always equate with influence, as Sherry Arnstein (1969), a US author on public participation, cautioned, noting:

> ... a critical difference between going through the empty ritual of participation and having the real power needed to affect the outcome of the process ... participation without redistribution of power is an empty and frustrating process for the powerless. It allows the powerholders to claim that all sides were considered, but makes it possible for only some of those sides to benefit. It maintains the status quo. (Arnstein 1969, p. 216)

When, in a subsequent research study (Bell and Sweeting 2013), I interviewed the women from the local environment group, discussed above, about the local consultation processes, they expressed a great deal of frustration, stating:

There's a lot of muddying the waters that goes on, so you can't really follow what's happening. They don't follow the processes they are supposed to or reply to things. You just end up getting confused and give up. (Tracey)

I've noticed a distinct increase in bullying, instead of calm discussion ... You end up feeling hurt and disgusted and demoralised. (Sue)

This is how a group of working-class people feel who are confident enough to get involved with planning processes and who have a local community group who can supply information and support. However, there are many more working-class people who do not have this confidence or such a network, as Sharon explained:

It's interesting how [people on the estate] *don't believe that they've got a say, ... don't feel like they have a right and, even if they knew that they had a right and you give people permission, they still don't know the systems and, even when they start dealing with the systems, a lot of the systems kind of punish them or take something away or watch them, so then they don't see it as being able to work with them or develop something with them. I don't think they know where to go to make the demands ... and they get brushed off a lot easier.* (Sharon)

Sharon referred to the benefits agency, social services, housing services or even community services in terms of them having the ability and potential to 'punish' people for not having the 'correct' point of view. This view is consistent with my experience in relation to the planning consultation described above. After that, my local environment group was no longer consulted on any local authority activity and some of us were ostracised by some of the Council officers and representatives involved (all of whom have now moved on to work in other areas).

These experiences are also reflected in the wider literature on community engagement and consultation which indicates that planning and other environmental decision-making processes often exclude working-class and other marginalised groups and tend to support agendas created and controlled by elite groups or those from outside the community (e.g. Boutelier et al. 2000). Some analysts have argued that most so-called participatory decision-making procedures are flawed and used to manipulate and legitimate unfair outcomes. For that reason, participation has been described as 'the new tyranny' (Cooke and Kothari 2001). People need to have both the opportunity and the necessary support to participate in making decisions. Sharon also made this point, stating:

> *People get put off because there's too many steps, you have to do this, you have to do that … Their lives can change so quickly because you haven't got any buffers, do you know what I mean? Life just changes so quickly … They're always caring for someone, you know, especially the women … I think they get put off more easily, let's put it that way. "Oh, it's too much!" Even trying to get a constitution together to go for the funding, "Oh, I can't be bothered with that". So,* [as far as being listened to], *they never even get to make the demand a lot of the time.* (Sharon)

Some blame working-class communities, themselves, for their lack of influence in local environmental decision-making processes, alleging that they are uninterested and apathetic. Professionals often describe such individuals or groups as 'hard to reach' in terms of getting them involved in their projects and consultations. This is sometimes justified by the 'post-material values thesis' (Inglehart 1990), which suggests that poorer people are too preoccupied with meeting their basic needs for food, warmth and security to be able to consider or respond to environmental issues (see Chap. 6). Sharon's point would seem to confirm this. However, sometimes dismissing groups as 'hard to reach' is a means of avoiding exploring alternative, and more accessible and appropriate, mechanisms for their inclusion and participation. Therefore, we should examine the capacity or competency of the professionals and their agencies when working-class communities are excluded. The working-class people I spoke to had rarely been invited to give their opinion on local developments or policies. For example, when I asked Dave, who had lived on a council estate most of his life, if he had ever been involved in any local authority consultation, he replied,

> *Quite honestly, we never heard from the Council. The only time you ever heard from the Council was when you hadn't paid your rent … So, no, I think the*

answer to that would be no, and if it had taken place, I wasn't aware of it. (Dave)

As discussed earlier, despite all these barriers, working-class people sometimes do get involved with environmental decision-making, often at great personal cost. It is then immensely frustrating for them when their expert knowledge of local issues is ignored. In my earlier research (Bell and Sweeting 2013), Deb and Sally, who had been involved in some Neighbourhood Renewal meetings, for example, said,

They seem to walk all over us or send us people who don't care who aren't from the area.... [We] say what we want and then it just gets chucked in the bin. (Deb)

... it seems that those who manage the community, they don't live here so they don't know the real problems we face and what really make us happy. They do what is good for them ... When we are asking for some of the things that might disturb their programme or agenda ... they don't go along with our request. (Sally)

Local people are often in the best position to know what is needed locally but many of my research interviewees felt that, when they lobbied to prevent or reduce environmental problems, they were often met with a closing of ranks among the elites. For example, local working-class people I interviewed who were campaigning to prevent an 'Incinerator Bottom Ash' (IBA) recycling plant in their local area described this frustration. The proposed plant was to use ash generated by 'Energy from Waste' facilities, where waste is burnt for heating. The incinerator bottom ash would then be processed to make an aggregate suitable for construction projects and the manufacture of asphalt (Days Group 2017). This would involve bringing the IBA to their local area for storage and processing. The proposed plant was controversial for a number of reasons. Creating energy from burning waste is highly problematic in terms of sustainability and some authorities have, therefore, stated that incineration should not be considered a renewable form of energy (see UK Select Committee on Environment, Transport and Regions 2001). The health and safety aspects of IBA have long been questioned (FOE 2002) with recent scientific studies warning of the potential toxicity (e.g. Dung et al. 2017). Paul, one of the local residents campaigning against the proposed plant, emphasised the potential health hazards:

> *What they wanna do is … dump 110,000 tons of it behind my house and then turn it into concrete blocks and road surfaces. Now, the whole predication for this business process is that it's inert and nontoxic [but] … it's not nontoxic and it's not inert by any means … It contains anything that goes through waste streams … and what you get in the bottom ash is cadmium, lead, all the heavy metals … you really don't want to be eating it, you don't want to be breathing it, you don't want to be drinking it … and they're sliding left, right and centre … to tell me that this stuff is perfectly fine, to store it 10 feet from somebody's house on a dock that's got 70mph winds, you know. I'm not stupid.* (Paul)

There was initially a permit refusal, but the aggregate supply company appealed the decision and there was a subsequent enquiry. Local people were dismayed that, just before the enquiry opened, the Environment Agency dropped its opposition to the granting of the permit and, at the end of the enquiry, the IBA plant was given planning permission (McKenna and Olswang 2018).

Paul felt that unfair tactics had been used by authorities and companies to prevent him and others from exposing and campaigning against this, and other, local environmental hazards:

> *I've been closed down left right and centre … I'm ostracised and victimised … [They] … don't like bad publicity … They are trying actively to just get rid of you … [They] … get people that are sympathetic into positions within the community action groups … and start up separate grass roots organisations that aren't actually grass roots organisations and then divert all the funding to them.* (Paul)

Such sentiments echo many studies of public participation (see e.g. Dargan 2004), with residents disappointed with the level, form and extent of response offered to them by public authorities. Many of the tactics the local authority and other officials had used to side-line Paul and his community were familiar to me from my experience with the environmental activist group on my estate (see Bell 2008) and my community development work over the years. This included bullying and discrediting tactics and the manipulation of funding to encourage 'alternative voices', that is, the ones that agree with the Council officers or the companies in question, as Paul described. Obviously, it is not local authority policy to punish working-class people for their views and most of the officers that work for local authorities do not, of course, behave like this and try their best to treat everyone with respect. However, I have seen it happen on many occasions.

While the UK Localism Act supposedly gave citizens more control over planning decisions (Localism Act 2011), many working-class communities are still unable to halt unwanted and health-impacting developments in their localities. Part of the problem is that this legislation gave the middle-classes an advantage because of their greater confidence, resources and networks, as illustrated in the next section on street cleanliness.

Street Cleanliness

National research shows a clear pattern of less street cleanliness in the most deprived areas in the UK (e.g. Bramley et al. 2012). For example, in Bristol, 49% of residents in wealthy Stoke Bishop think that street litter is a problem (BCC 2017d), compared to 80% in Lawrence Hill, a deprived part of the city (BCC 2017e). This may seem a relatively trivial issue, as it does not directly kill people (although by attracting vermin, it is a health risk). Yet it is an indignity and an injustice that working-class people have to experience dirtier and less attractive streets and public spaces, impacting on self-esteem and sense of wellbeing, as well as the local economy. It can also reinforce the decline of working-class areas, as Amy and Lauren describe here:

[The streets], *they're dirty. I don't know how often they clean them but not enough … The streets were dirty all through the hot summer … They* [the Council] *hadn't put it into their budget … but they should have done because, obviously, tourism is a big deal in this town. There's not a lot else really going on so it's important to attract visitors to come back.* (Amy)

It is depressing when you go outside sometimes. You think, I better go out to get some exercise and fresh air and then you see all the rubbish everywhere, loads of broken stuff, things falling apart. You come back feeling worse, depressed, like you don't want to live round here at all … It didn't use to be like that. Council estates were nice places. Especially the last five years it is awful—it is like a rubbish tip these days. (Kelly)

Policy on street cleaning has historically tended to emphasise individual behaviour, blaming 'litter-lout' residents, and has sought to change this through education, persuasion and punishment (Lewis et al. 2009). However, in the last decade or so, there has been recognition of the social characteristics of working-class neighbourhoods which could lead to the need for a particular form of service provision which is appropriate to the local context. Hastings et al. (2005) argued that there is more litter and waste created in poor areas because people spend more time at home or in the neighbourhood; there are greater numbers of children; people buy cheap goods which have to be replaced and thrown out more often; dwellings often have more occupants; there is a greater use of cheaper food which tends to be heavily packaged; and there is less car access to take waste items to civic amenity sites. Sometimes such areas can be littered by people passing through who perceive the area as of little value. If the litter is not then removed, this increases the chances of other littering. A number of studies indicate that street cleaning services in the UK are sometimes developed without considering these different needs (e.g. Hastings 2007, 2009; Hastings et al. 2014; Bramley et al. 2012).

For example, Hastings (2009) and Hastings et al. (2014) looked at three local authority areas finding that, while they all cleaned residential streets on a twice-weekly basis, two did not take into account the differing level of need in the more working-class areas. When the researchers asked the street cleaning manager in the first area if needier areas could receive an augmented service, the manager replied, 'It would be politically unacceptable to be seen to be favouring one part of the population over another' (Hastings et al. 2014, p. 212). He felt it important to provide a standardised service, regardless of variations in need, so as not to attract criticism (presumably from middle-class householders). In the second local authority area, the street cleaning service did take account of greater need in the working-class neighbourhoods. However, the interviews with managers indicated that the targeting had to be done by stealth. The official version was of a 'one-size-fits-all' standardised service, but the work was distributed so that there was extra capacity in areas with higher levels of need—predominantly the working-class areas. In the third local authority, 20% more was spent on routine street cleaning in the most affluent areas than in the most deprived areas. The street cleaning managers explained skewing resource and capacity to better off areas in terms of it being politically and managerially expedient. They preferred to allocate extra resources to cleaning the streets in the wealthier areas so as to minimise complaints, since it was the householders in these middle-class areas that were more likely to voice their objections to an inadequate service. The researchers called this 'middle-class capture' to explain how the middle-classes tend to be the main beneficiaries of this public service.

Middle-class capture of public services has been a long-recognised phenomenon. In 1982, Le Grand noted: 'Almost all public expenditure on the social services in Britain benefits the better off to a greater extent than the poor. … In all the relevant areas, there persist substantial inequalities in public expenditure, in use, in opportunity, in access and in outcomes' (Le Grand 1982, pp. 3–4). Although this statement is now considered to be an overgeneralisation, there is some consensus that middle-class groups are advantaged with regard to some public services in particular places (Hastings et al. 2014). The way that 'middle-class capture' occurs in relation to environmentalism, more generally, will be further discussed in Chaps. 5 and 7.

Waste Collection Services

Whilst waste collection services have improved dramatically in recent years, insufficient attention has been given to equity and social considerations. Expectations placed on local authorities in waste and recycling have increased markedly, in particular, since the passing of the EU Landfill Directive. Since its conversion into UK waste policy, local authorities have been required to hit increasingly demanding targets on the reduction of waste going to landfill.

Waste reduction has been discussed in terms of four options: reduce, reuse, repair and recycle, but the policy framework in the UK and most other countries favours recycling over all the other options, despite it having the most negative environmental impact of the four. If we look at the example of packaging, we can see how this emphasis works for the benefit of business at the expense of householders. It is estimated that one quarter to one-third of domestic waste is made up of packaging (The Open University 2008). Yet businesses are largely unwilling to reduce their packaging because it is used to promote their brand and an associated lifestyle. Unfortunately, this means householders are carrying the burden of this focus on recycling: a burden which particularly impacts working-class people who can struggle to sort and clean their waste because of 'time poverty'. Those on lower incomes tend to have less free time as a result of, for example, greater caring responsibilities, or longer hours of work (Burchardt 2008). Section 46 of the Environmental Protection Act 1990 enables local authorities to impose penalties on householders who present their waste incorrectly for collection. There is no consideration that some people genuinely find it difficult to meet the obligations imposed on them and little treatment of people in any socially differentiated way.

In earlier research, David Sweeting and I analysed household waste and recycling (Bell and Sweeting 2013). We found that the working-class residents were frustrated with the service and felt burdened by the requirements. Some of them pointed to the underlying issues that create waste problems in the first place, as illustrated here in Paula and Junes' comments:

Things now break down much quicker … When it's just outside the warranty, usually, that's when it breaks down [laughs] *…so you have to keep buying more.* (Paula)

It's putting everything on the individual. I mean, why should we be messing around with waste? The Council don't have to do so much now as we have to spend time sorting it and, it's like with the electric, before they used to come and read the meters, they employed people to do that. Now they've laid them all off and we have to do it ourselves. (June)

The working-class people I spoke to in the research for this book also complained about the cost and inconvenience of the waste collection services. Although some services are free at the point of use, others incur charges. 'Green waste' or garden waste had previously been routinely collected as part of normal collection procedures. However, in most areas, residents are no longer allowed to dispose of the waste in the landfill container. This is positive in terms of the environment, as green waste produces methane gas and so needs to be segregated, but in many parts of the country residents now have to pay for its removal and recycling. In Bristol, for example, a garden waste bin is available to householders for an annual charge of £32, plus an initial charge of £21. Though there are concessions for some benefit recipients, this is unaffordable for the working poor.

This trend is reflected nationally with more than half of councils now having introduced charges for garden waste collection, previously paid for through their council tax. The Local Government Association has said that councils are forced to charge for this as an extra because they face a £5 billion shortfall in funding from central government.

Some residents who cannot afford the fee respond to the build-up of green waste by burning it in their gardens. This impacts health and causes anxiety as some residents described in our earlier research (Bell and Sweeting 2013):

On a Sunday night it's like Beirut with all the garden waste being burnt so it is a big issue, you know, how to get rid of it. (Paula)

There's fires all through the summer … one day they had to get the fire brigade out. People are burning the waste because they can't afford to get it collected … If you ring up and complain about the fires they want to know the address but then the neighbour gets in trouble, maybe a fine … That does not solve the problem because nearly everyone is doing it. (Tracey)

There are also similar problems caused by a requirement for payment for the collection and disposal of large household items (e.g. furniture). Working-class residents without either the access to a car to enable them to take large items to the municipal dump or the means to pay for the collection of large items are disadvantaged by charging for collection services. There were feelings of frustration and unfairness about this, as well as about other aspects of the service, as exemplified in the comments made by Joan, Irene and Amy about their respective local Councils:

> *I saw an old armchair dumped in the lane next to my house, so I brought it in and rung up the Council to get it taken away. It took me three days to get through and £11 in phone calls and then they say it would be £25 to take it. It's not even mine—I was just trying to help keep the place clean. Now it's stuck in my garden. The Council won't spend the money to keep the place clean, they are cutting everything so now there is junk everywhere.* (Joan)

> *It's hard for people to pay to get big items taken away. If they're not on benefits, you've gotta pay … I know a girl that actually wanted to get rid of a fridge freezer and they said, oh, we'll get back to you with a date, a time. Nobody got back to her. She put it outside her house and it was there for over a month … She was waiting* [for it to be collected] *and, in the meantime … the glass shelves were out, they were smashed on the pavement.* (Irene)

> *Glass, textiles or anything like that, you've got to take it somewhere. It's usually out of town so, unless you have transport, it's difficult. They did used to allow people a garden waste bin but now you have to pay £45 a year for that, which a lot of people really complained about … There are lots of things they refuse to take altogether, so you have to get to the household disposal place … For us, from here, depending on the traffic situation in the town, it could take hours.* (Amy)

Several reports (e.g. Lucas et al. 2004, p. 8) have noted that fly-tipping may be the result of, or at least aggravated by, local policies and national legislation that require people to pay for disposing of certain kinds of waste. In the current political and economic climate, waste collection services are potential targets for contracting out, possibly driving down the quality of service. Educated, middle-class groups tend to be more likely to be able to influence waste service delivery to favour their own area (Hastings 2009), exemplifying, again, the process of 'middle-class capture'.

As discussed in the last chapter, working-class people not only struggle to deal with the demands of waste management regarding their own and local waste, but often carry the burden of waste management for the wider nation or world when it is brought to their communities for burning, breaking and storing, as Paul highlighted:

A lot of the waste industry is seen as a green thing when, in reality, it's not. It could be, you know, quite a lot of it could be good stuff to do, but just not right in the centre of people's houses and that's the issue for me. You know, industry happens, and we've got to deal with our waste, but the issue is that, currently ... it's all in [local area]. *Anything you don't want anywhere else, stick it in* [local area]. (Paul)

Consistent with this comment, a new energy-from-waste facility is now being constructed in the deprived area where Paul lives. The plant, to open in 2020, will burn up to 320,000 tonnes of waste per year. This is a reflection of a national trend whereby the amount of waste managed by local authorities and sent to incinerators, or energy-from-waste plants, tripled between 2010–11 and 2016–17. If this trend continues, the amount of waste incinerated will overtake the amount sent for recycling by the end of the current financial year, according to a recent report by the Green Party (Green Party 2018). This means that much of the plastics, cardboard and paper that people clean and separate for recycling will eventually end up being burnt, together with all the other waste. As the Green Party report (2018) stated, 'There is a logic to generating energy from the waste that we cannot recycle, or reuse, but it is meant to be the last resort option. What we have created instead is a market driven system of incinerators which constantly need to be fed' (Jones in Green Party 2018, p. 1).

Ange, one of the residents living in an area where an energy-from-waste plant is being built, queried the need to bring waste to her neighbourhood:

I heard the rubbish that come to [local area] *has come from London. Why do they put the rubbish on the train and bring it here? There could be anything in it. If it is safe, why don't they burn it in London? Why do they bring it here where we live?* (Ange)

Sometimes the waste goes even further afield, to working-class and low-income communities in other countries. In order to reach UK and other high-income country environmental targets, environmental problems are often outsourced, that is, sent to far away low- and middle-income countries, and this is particularly the case with waste management and recycling. In particular, there is a problem with electronic and other hazardous waste being sent to the Global South. Since China has stopped the importation of 24 kinds of solid waste, including plastic, alternative destinations are being sought, and concerns are being raised that much of the waste that is being sent abroad for recycling is ending up in landfill or incineration in countries including Turkey, Malaysia and Poland (National Audit Office—UK 2018).

Waste management will never solve the problem of waste pollution because it is simply too huge. It has to be addressed at its source—by reducing unnecessary production. Some mixed waste is difficult and/or not profitable to recycle. The global waste management market is highly lucrative and seeing strong growth, expected to reach a value of $484.9 billion by 2025 (Allied Analytics 2018). Yet prices paid for recyclable material fluctuate daily, and with the glut of recyclable material following China's new policy, prices have dropped. Already around 70% of potentially recyclable plastic in Europe is sent to landfill or incinerated rising to about 90% in the US (e.g. Staub 2018) and the proportion rises when it becomes less profitable. We generally need much less plastic to be made. Recent bans on single-use plastic at local and national levels around the world are a great start.

By focusing on recycling and incineration, rather than reducing waste, especially packaging, the burden of waste management currently falls on poor and working-class communities, locally and globally, rather than the companies that produce the waste. These companies benefit from the advertising carried on their packaging as well as from the extra sales resulting from the continual need to replace their short-life products. They profit from waste management policies which focus largely on recycling, rather than the three steps that should come before that—reducing, reusing and repairing. Unless these three steps are prioritised, unnecessary waste will still have to be managed, burdening working-class people through costs, time and toxicity. When recycling becomes a problem for working-class people it threatens to alienate a large section of the population from environmentalism.

Pollution Policy

Another area of environmental policy that has a disproportional impact on working-class people is pollution policy. As I write today (23 October 2017), it was reported that London Mayor, Sadiq Khan, has just launched a £10 toxicity charge (T-Charge) aimed at older, more polluting vehicles on London roads. The charge is to apply mainly to diesel and petrol vehicles registered before 2006 and will operate in addition to (and during the same operating times as) the Congestion Charge (Monday to Friday 7 am–6 pm). This means that motorists will now have to pay £21.50 to drive in central London with an older car. This cost will hit those on lower incomes hardest, who cannot afford to update their car every year or, sometimes even, every decade. This will include delivery drivers, builders and others who need to transport their tools, and people on multiple casual contracts, having to move from one job site to another quickly. If people are driving old vehicles into London, already paying the congestion charge and parking costs, they will probably have essential reasons for doing so and have looked at other options and not found them workable. Many working-class people have already been forced out of living in central London by soaring house prices and so have to drive in to London to their jobs.

A diesel scrappage scheme is on offer, handing out £3000 grants so people can buy a new car, but this is essentially a subsidy to the better off who can afford the new car. Scrapping cars that are functioning and using resources to build new ones does not even seem environmentally sensible. What's more, recent research is showing more pollutants are caused by particles emitted from tyres and brakes (even new and electric cars have these emissions), construction and wood-burning than car exhaust pipes. For example, wood-burning stoves, which have long been marketed to environmentally conscious middle-class people (at a cost of as much as £1500) as a form of green heating, have been found to contribute to around 23% and 31% of particle matter (PM) emissions during peak times in some UK cities (Font and Fuller 2017). On one day in London in January 2017, the contribution of wood-burning stoves to the total PM pollution was found to be more than 50%. Some of the smaller particles, known as PM2.5, cause and exacerbate lung and heart conditions and increase the likelihood of heart attacks and strokes. Key researchers in this field have said that the harm far exceeds that of traffic pollution since people are exposed to high levels of traffic pollution mostly when they are travelling on busy streets for short periods, whereas wood burning pollutes residential areas and homes (e.g. Font and Fuller 2017).

Although the London Mayor also has plans to address the pollution from wood-burners and construction in the longer term, there has been much less focus on these issues. Wood-burners cannot be considered necessary as there are many cheaper alternatives for keeping warm. Although many were bought in good faith as environmentally friendly solutions to heating the home, they have become a status symbol of sorts. The T-Charge, being an extra cost to the working-class, not having a major impact on the causes of pollution and being enacted while middle-class people can still continue to pollute (with their new cars, wood-burners, newly built investment properties in central London, etc.), really sends out the wrong message to working-class people. Such a scheme is not only more likely to make life more difficult for working-class people but can also alienate them from environmentalism, as it feeds into the view that it is a luxury consideration that only benefits the middle-class. Although it is really important to reduce air pollution, which itself impacts more on the working-class, the policy has not been adjusted to make it acceptable and helpful to them.

Emmanuel Macron's defeat by the *gilets jaunes* (yellow vests) over a proposed green fuel tax could be seen as a warning to governments who try to impose such taxes on people who are struggling to make ends meet. Many advocate carbon taxes as a way to reduce the use of fossil fuels, with some considering them to be essential to avoid more severe climate disturbance (e.g. Stiglitz et al. 2018). However, when this occurs within a situation of social injustice, and when the tax exacerbates that injustice, it can be perceived as unfair and can have the damaging impact of alienating people from environmentalism. Research shows that people tend to reject carbon taxes because they perceive that they do not work and that their real purpose is not to reduce greenhouse gases but to increase government revenues (Carattini et al. 2018). Among the working-class interviewees for this book, there was little support for green fuel taxes, and similar suspicions about motives and justice. For example, Julie said:

> *They talk about taxes on petrol to stop car emissions but then you have got all these big companies that pay off whoever and they are spewing out all this stuff … going into the oceans and into the air… We have to pay for what little we do, but they are still spewing it into the air because they can get away with it … I think they [the government] rake in enough taxes anyway that they should have enough money to put it on renewable things. You think where is it all going anyway? You don't seem to see where all the taxes go. (Julie)*

Protecting low-income groups from the regressive aspects of green fuel taxation tends to increase public acceptability, as does ring fencing of the funds raised for social or ecological progress, the availability of alternative transport for those who cannot afford the tax and trust in the government proposing the tax (Carattini et al. 2018).

The fuel tax imposed in France that led to the initial *gilets jaunes* protests could have been replaced by policies to increase income taxes on the wealthy combined with investment in reliable public transport for the same environmental benefit whilst also supporting those on low incomes. Working-class people are not likely to accept policies that put additional pressure on individuals, especially the least well-off. While cars need to be taken off the road, this should be part of a planned strategy to provide alternatives, which would improve living standards and could increase jobs (see Chap. 8).

Energy Policy

Linked to the issue of pollution is the question of energy policy. A number of academics have recently highlighted the importance of considering 'energy justice' for effective and informed environmental policy making (e.g. McCauley et al. 2013; Sovacool and Dworkin 2015; Jenkins 2018). For example, Sovacool and Dworkin (2015) advise considering energy policy in terms of availability, affordability, human rights, good governance, sustainability, intra-generational equity, and intergenerational equity and responsibility. This requires balancing the different aspects simultaneously, an approach which seems to be currently lacking in the UK to date.

Energy policy in the UK is driven by the UK's legally binding commitment, set out in the Climate Change Act 2008, to an 80% reduction in greenhouse gas (GHG) emissions by 2050 against a 1990 baseline. This Act has been generally welcomed and is certainly necessary but, currently, the policies that support it can make life more difficult for working-class people. UK government programmes to reduce CO_2 emissions are still funded by consumers via energy bills—that is, all customers pay a set amount on each unit of energy consumed. This has a greater impact on poorer people, whose energy bills make up a larger proportion of their household expenditure. The UK Conservative government's Energy Act, 2011, has rarely been helpful for those on low incomes (Stockton and Campbell 2011; Ekins and Lockwood 2011). It includes finance arrangements to allow householders to pay for energy efficiency measures through loans attached to their property, to be repaid through their electricity bill. Many working-class people are unable to access loans or are unwilling to take them out because of fear of debt. In addition, many live in rented properties, so they would not be able to make the necessary alterations to their homes.

Another UK energy policy, Feed-in Tariffs (FiTs), similarly favour the better off. Under FiTs, an obligation is imposed on regional or national utility companies to buy the energy derived from renewable or low-energy sources. Those who have enough income to invest in the equipment can install microgenerators and sell the excess electricity generated. FiTs, therefore, provide economic opportunities for wealthier people, that is, mostly middle-class, who can afford to invest in the new technology. Again, the programme is funded through an additional levy on all consumer bills. Energy costs are a key driver of fuel poverty (The Committee on Climate Change 2017). Consequently, there is, not only injustice with regard to costs falling disproportionately on low-income groups, whose energy costs make up a larger proportion of their financial outgoings, but there is also a lack of equal access to the potential benefits of this policy.

However, the costs of renewables are a relatively minor reason for the price hikes in energy bills. The social and environmental policies that energy suppliers are required to deliver on behalf of government ('obligation costs') are only about 5–15% (Competition and Markets Authority 2016). Rising prices are mainly driven by the profit-maximising business model of the utility companies.

The concept of 'energy justice' incorporates consideration of energy production and consumption. It was originally conceived to cover 'infrastructure siting, (e.g. wind farms, nuclear waste facilities, etc.), subsidies (e.g. renewables, nuclear energy), pricing (e.g. fuel poverty) and consumption indicators (e.g. smart meters) within the context of global and local pressures' (McCauley et al. 2013, p. 108). In my experience of attending environmental community events, environmentalists do sometimes discuss 'energy justice' in this sense, linking the desire for sustainable energy with energy security and energy equity. However, their narrative mostly focuses on developing small-scale community energy schemes, rather than challenging the costs of energy charged by the six large energy firms which currently supply about 80% of UK domestic customers. Focusing on the energy technology is not going to interest working-class people very much when they are struggling to pay for it. As Liam said,

I am working as a union rep and what people that live in poor communities are saying that they are worried about is paying for their energy, not the source of the energy. (Liam)

The privatisation of UK energy has been problematic for working-class people living on low incomes. Overall, energy prices have been climbing dramatically since privatisation, with average domestic electricity prices rising by around 75% between 2004 and 2014, and average domestic gas prices rising by around 125% over the same period (Competition and Markets Authority 2016, p. 5). Ofgem have criticised the price rises as 'hard to justify' (in Vaughan 2017). While 30% of UK consumers are worried about paying their energy bills (Ofgem 2017a), the six big energy companies—British Gas, EDF, E.ON, Npower, Scottish Power and SSE—have made over a £1 billion in profits over the last year. Several of the working-class people I interviewed for this book commented on the anxiety caused by the high energy costs. For example, Dave remarked,

> ... *it causes stress because you're always balancing one thing you're paying off against something else, and these* [energy] *firms are pretty merciless in getting their money back, you know. I mean alright, you could say on one level the worst thing they can do is send you to prison, and then you'll have free meals* [laughter] *and free heating! But, I think that's a pretty drastic outcome to the situation, which I would rather avoid, yeah* [laughs]. (Dave)

Privatisation was touted as enabling consumers to keep their energy costs down because of the ability to switch suppliers. However, 60% of people do not switch because it is complicated and often ties up significant amounts of money with the previous supplier for weeks. As discussed in Chap. 3, it is working-class people that are the least likely to switch (Ofgem 2012). Ofgem explain: 'Nearly half of consumers who are unemployed, or have intermittent, semi- or low-skilled work, have never switched supplier, compared to under a third of other consumers. This means that they are more likely to be on expensive standard variable tariffs, despite being less able to afford them' (Ofgem 2017a, p. 7).

These examples illustrate how ill thought through energy policy can exacerbate the difficulties faced by working-class people. Though renewables subsidies are only a small part of the problem, they could be viewed as the main problem, potentially alienating working-class people from environmentalism. If prices were capped and companies forced to cut profits, this would make a significant difference to working-class people. More radical, but probably more helpful, would be, as the UK Labour Party manifesto of 2017 called for, a 'transition to a publicly owned, decentralised energy system'.

Green Jobs

A 'green job' has been defined as 'work in agriculture, industry, services and administration that contributes to preserving or restoring the quality of the environment' (UNEP 2008, p. 53). Green jobs are often seen as a way of simultaneously addressing social and environmental issues, reversing deindustrialisation, diversifying economies and bringing new employment opportunities for a skilled workforce. This should all benefit working-class people, but does it?

Green jobs are portrayed as benefiting working-class people in part because they may increase the number of jobs overall. The UK Renewable Energy policy has explicitly claimed that investing in energy renewables will bring thousands of 'green jobs' (in Deutz 2014). Related policies, such as the Low Carbon Transition Plan (2009) and the Low Carbon Industrial Policy (2009), also mention jobs in relation to: low-carbon vehicles; offshore wind; carbon capture and storage; wave and tidal power; nuclear power; and low-carbon and renewable materials and manufacturing. The UK government has estimated that 70,000 new jobs could come from investment in offshore wind alone (DECC 2009).

Studies in the US (e.g. Wei et al. 2010) and China (e.g. Cai et al. 2011) suggest that these claims may be correct, since they have found that renewables create more employment than the fossil fuel industries. However, whether this always occurs is controversial and may depend on the sector and the location. The renewable industry is still constrained by the same drives to reduce costs and so it does not follow that the renewable jobs created will be in the same places where the dirty jobs have been lost. If it is cheaper for the company to move, it will.

Helm (2017) considers that some regions of the world will face particularly adverse impacts from a low-carbon transition, such as the Middle East and Russia, due to the importance of oil and gas for their economies. Europe and the US may do better since they can compete in the supply of new energy sources.

Questions have arisen regarding the help given to people employed in the polluting industries to transition to these new jobs. Are the jobs available to manual working-class people or do they require Higher Education training to access them? Who, in fact, gets these jobs? In the US, Mason (2009) found that Black people, Latin Americans and women had difficulty gaining employment in green sectors. In general, there are few good examples of where there has been help for workers to transition to green jobs. However, the German government, during a decline in the production of coal in the 1990s, created widespread retraining programmes to help coal workers find new employment, often in the renewable sector (Miller et al. 2013). There are also other examples, globally (see Gambhir et al. 2018).

The working-class people I interviewed in the course of my research were quite sceptical about green jobs, if they had heard of them at all. One woman complained about green jobs not going to local people, even when they are sited in, or near, disadvantaged parts of the city. She felt that the workers and customers of the local green companies seemed to look down on the working-class people living nearby:

> *In* [local area], *the Green Business Park* [visitors] ... *had been getting lost on the estate and there's been quite a few coming in and they've been asking how they get back to the ... Green Business centre. When they ask, "they don't look us in the eye", people I know in* [local area] *say, it's as if they're not talking to them as another human being ... When you say, oh this is going to bring jobs, but is it going to bring jobs to people in* [local area]? *Please! ... They're not jobs for people in* [local area], *that's for sure.* (Sharon)

Sharon is referring to a Green Business Park which was promoted as providing around 350 jobs for local people in one of the most disadvantaged areas of the UK. The current tenants are quite diverse though many provide services which would require professional employees and some cater to wealthier consumers, such as the organic cold-pressed chocolate company, the 'boutique' dentist and the accountancy firm. I could obtain no information or find any record of how many jobs actually went to local working-class people in the course of carrying out research for the book. It seems to be the local perception that there were few.

Even when 'green jobs' are available to working-class people, they are not necessarily healthy, decent or well paid. UNEP (2008, p. 53) states that green jobs should 'meet longstanding demands and goals of the labour movement, i.e. adequate wages, safe working conditions, and worker rights, including the right to organize labour unions'. The word 'adequate' is problematic for some, as it does not suggest well paid. Even in high-income countries, green jobs tend to be relatively low paid and people may have precarious employment contracts (UNEP 2008). Green businesses face the same pressures as other capitalist enterprises to lower prices in order to sell more and to make a profit for their shareholders (Deutz 2014). Recycling jobs, in particular, are often dirty and dangerous (UNEP 2008). There are also questions regarding the displacement of green jobs, such as recycling, to low-income countries where it may be more difficult to monitor and control toxic processes and lax worker protection (Stevis 2013). We need to take into account a range of factors in making our assessments of green jobs, including investigating the supply chains.

Sometimes, so-called green jobs are not even green in the sense of being sustainable in a holistic way, such as jobs in the nuclear power industry. While creating less greenhouse gasses as a direct by-product, nuclear energy still creates toxic and dangerous emissions, in addition to toxic waste which takes thousands of years to become safe, as well as posing a high possibility of accidents (see, for example, Wheatley et al. 2016). Some argue, therefore, that 'green jobs' are part and parcel of 'greenwashing'—creating an illusion of environmentalism without having to make the real changes needed. This was also the view among some of the working-class people that I spoke to. For example, Dave said,

> I don't wish to sound cynical but, I've just got this feeling again that there seems to be these superficial remedies … When we look deeper at the problem, underneath nothing gets actually resolved. I'm not against those [green] jobs in principle, what I'm against is … these things sew a false illusion, or a false solution to a more serious underlying problem, you know. (Dave)

So, although campaigns for green jobs could unify environmentalists and workers, unless they are genuinely green and genuinely social, they can be just another source of working-class cynicism about environmentalism.

Green Programmes

Another key way that working-class people are excluded or alienated from environmentalism is through the environmental programmes that attempt to engage them. I will illustrate this point with an example of the

Bristol Green Capital project. In 2015, Bristol won this European award, supposedly in recognition of its environmental achievements, future commitments to sustainability and ability to inspire others to embark on a similar green trajectory. The Bristol Green Capital project aimed to create 'a low carbon city with a high quality of life for all' (Bristol Green Capital Partnership 2014). Bristol has many features that support it to achieve this goal. However, despite its green image and being voted 'best city to live in the UK' in numerous charts, like many other cities in the UK and globally, it is economically unequal and socially divided. That year 16% of residents were living in areas designated as the most deprived wards in the country. There is a wide range of deprivation across the city with 49% of residents in one area (Fulford Road North, Hartcliffe) experiencing income deprivation (Bristol City Council 2015). In addition, 22.6% of children under 16 were from low-income families, significantly more than the England average, rising to 45% in the most deprived ward (Bristol Health and Adult Social Care Scrutiny Commission 2015). The City Council itself recognises that low-income and other disadvantaged groups in Bristol are more likely to experience environmental deprivation, for example, in terms of access to clean air; healthy food; energy for heating; quality green space; affordable transport and safe streets (Bristol City Council 2015). These factors result in poorer health outcomes so that there is a persistent and significant differential in life expectancy between the most and least deprived areas of the city at 9.6 years between the highest (Henleaze) and lowest (Southmead). Impairment-free life expectancy shows an even greater gap (Bristol Health and Adult Social Care Scrutiny Commission 2015).

European Green Capital 2015 status presented an important opportunity for Bristol to develop an inclusive green transition process which could have simultaneously addressed some of these health and wellbeing disparities and been a model for the world. Shortly after gaining the award, an 'Inclusion Group' was set up, ostensibly to ensure that the programme of activities would represent the interests of, and cater to the needs of, diverse social groups. I joined this group on my own initiative, as a member of the Lockleaze Environment Group. In the next chapter, I write about how this group functioned, since it is more relevant to a discussion about environmental organisations. Here, I will focus on the City

Council's programme and policies. Firstly, it was reported to us in the first meeting that there was to be no financial support from the Council to help make the Inclusion Group accessible. Since those living on the predominantly working-class outer estates were having to pay at least £5 return bus fare per meeting, as well as give up hours of their time, they did not attend the initial meetings. After repeatedly asking for access costs to be paid, and being refused, I ended up paying their expenses myself to enable them to come. Much later access costs from the Council were forthcoming but it was too late by then for the people from the outer estates to have much influence as much had been decided.

Later, some of those interested in equality in the Inclusion Group collaborated in a funding bid to the 'Green Capital Strategic Grant' to help amplify the voices of equalities groups in the city on environmental issues, including working-class people. However, the project was not selected for funding while short-term, non-participative art projects, in my opinion unlikely to attract the interest of working-class people, seemed to be well funded. For example, £37,000 was given to a project designed to produce orchestral music to the sound of nuts falling from a tree (Telegraph 2015). I am not sure who was intended to benefit from this project but, ultimately, no one benefitted because it transpired that the tree in question was dormant in the Green Capital year. No nuts fell that year and so no orchestral notes were heard, evoking a mixture of amusement and anger among local working-class people.

Almost all the working-class people I talked to about Green Capital were negative about it or had not heard about it at all. A common response when I asked people about it recently was, 'When was that?' Of those that had heard of it, their primary concerns were that a lot of money was spent with little result and that few practical, lasting and worthwhile changes had occurred. Some working-class projects did get funded but, as explained here, considerable barriers had to be overcome to do this, in terms of publicity, culture and access:

Even when things were advertised, the information was on the internet and a lot of people don't use the internet … The people sat around the table at meetings were used to bidding, were used to being a part of that formal structure, if you like, so it was difficult to be yourself and be different … because of the words you use, the way you act … They didn't have money for transport to start off with. These meetings are never or hardly ever on the actual estate. Some couldn't go because of childcare … children aren't always welcome because it's very formal. (Anonymised to protect the identity of the participant)

This participant also talked about the resentment that was felt towards middle-class environmental groups and activists trying to connect with community groups and individuals in working-class areas so that they could submit funding bids which had an inclusion aspect to them. This outreach may have been for altruistic reasons, but some working-class people perceived this in terms of middle-class people using them to increase the likelihood of gaining funding for themselves or their organisation. For example, I heard that

Projects have happened [on the estates] *but it was like things had been done to them, rather than with them and there was quite a lot of resentment about that. People were getting money off their backs. The amount of times I saw that happening, poverty pimping … They were just putting in claims, putting in bids* [to work with people on the estates] *without even the people on the estates knowing about it. I saw that happen.* (Anonymised)

Another of my working-class interviewees who had also joined the Inclusion Group, but then left in frustration after a few meetings, said:

When Green Capital was starting to happen, I was concerned that it would be a lot of corporate green washing, rather than anything tangible that would benefit the city residents and, by that, I mean benefit all of the city residents. Also, I was concerned that it would be a very patronising … talking down to the majority of the residents in the city, telling them how they should live their lives and it would be very lifestyle based … [so] *… I invited myself to the Inclusion Group meetings … I found it very annoying … I remember there being conversations, even about where some of those mingle events could happen … and they're all happening mostly in the city centre in slightly expensive cafes and bars and stuff.* (Anonymised)

Like most of the other Bristol research participants, she also felt that the funding had not been distributed well, in terms of lasting value or legacy:

What I saw was piecemeal bits of money going to little groups but nothing that would enable their longevity or their growth or expansion. There was a lot of conversation about legacy, what would the legacy be? I don't see a legacy … I saw a lot of billboard posters. I saw adverts on buses, but I didn't see much real tangible change in the city or a commitment to tackling some of the environmental issues that are here in this city…nothing very meaningful and I doubt had any real impact on the lives of the majority of residents of the city.
(Anonymised)

After about a year of involvement in the Inclusion Group, I also left as I did not feel that we were working on the agenda that I had hoped, I felt I was not being listened to and it seemed that the meetings were dominated by middle-class people and the agendas they wished to focus on. Since the Green Capital year passed, there has been an official weighing up of its benefits. On the positive side a BAME organisation was set up, 'Green and Black Ambassadors', which emerged from Ujima Radio's involvement with the Bristol Green Capital Partnership (BGCP). This is an excellent initiative which promotes and celebrates BAME involvement in sustainability and environmentalism. There also seems to have been quite a lot of work happening in the schools which would have included working-class children and may perhaps lead to some legacy in terms of attitudes. For example, one of the interviewees for this book said:

I didn't have much to do with the Green Capital, but my son did. He was involved with that and he got a Blue Peter badge because of it. To be honest I didn't pay that much attention to it. I genuinely can't remember much about it.
(Anonymised)

On the more negative side of the weighing up, Steven Williams, an ex-Bristol MP, has been trying to find out how the public funds invested in Green Capital were spent: £8.3 million of taxpayers' money (made up of £7 million from the national government and £1.3 million from local government). He has not been able to get answers to most of his questions (Williams 2016).

The new Mayor, Marvin Rees, set up an independent review of the city's year as European Green Capital, conducted by Steve Bundred of the Local Government Association. His report (Bundred 2016) explicitly stated that the Council should have given more thought and devoted more resources to addressing the barriers that prevent members of excluded groups from attending meetings or engaging with the programme. He said what was required was 'a paradigm shift within BCC' (p. 5). He criticised the council for not compensating deprived people for loss of earnings and personal costs for getting involved with the meetings surrounding the programmes, and added,

> ... a few more events away from the city centre might have made attendance at them more attractive to traditionally excluded communities ... there was little evidence of deep understanding of the links between social justice and environmental issues such as the need to provide the economic conditions needed to permit poor people to participate. (p. 24)

He also criticised the Council for focusing too much on getting people to turn up to events and not enough on involving disadvantaged groups in setting the agenda. There was also criticism of the Green Capital grants programme for having 'an in-built bias in favour of existing groups and "insiders"', recognising that support was needed to assist community organisations not used to submitting grant applications to navigate their way through the system. Bundred also explicitly said,

> ... the Council could have done more to demonstrate its commitment to reaching out to all parts of the city by responding with greater vigour to a pollution problem in Avonmouth during the Green Capital year. (2016, p. 24)

However, despite all these critical comments about Green Capital, the Bundred report went on to minimise responsibility, stating that these failures should be seen as 'a few missed opportunities identified only with the benefit of hindsight. Both the Council and the Company had a role to play and so too did others in leadership positions within the city. But all were well-intentioned. No groups or communities were deliberately excluded' (2016, p. 24).

Many working-class people that got involved with Green Capital did see it as deliberate exclusion to the extent that they were not listened to when they warned that the planned programme would not benefit them. They did not feel that the key organisers cared about them or the environment. For example, the following comments were made in relation to Green Capital (anonymised so as not to reveal the identities of the participants):

What a wasted opportunity! ... it was an opportunity for Bristol to make a difference to the poorest communities and we've got a lot of poor communities in Lawrence Western, Henbury, Avonmouth, Hartcliffe, that could really benefit with some proper investment and some proper sustainable investment and [the money] *was squandered on electric balloons, trees that didn't play nuts and... it's disgusting really ... and we're still paying for it!* (Anonymised)

What I've seen is ... no improvement in air quality in the city ... We're seeing another attempt to put some kind of incinerator in Lawrence Hill/Barton hill area. Again, massively urban environment, predominantly working-class people and people of colour, so we didn't see any local government changes that would say these things can never happen in this city ... I saw no voice for that. I saw no honest acceptance of those issues. What I saw is loads of celebration of stuff ... a lot of patting ourselves on the back rather than holding ourselves to account and saying what are we going to ensure for our future generation? (Anonymised)

Despite the Bundred report's recommendation for a paradigm shift, the remaining institutions of Green Capital are continuing to make similar mistakes. For example, in 2017, Bristol Green Capital Partnership (BGCP) launched a new grant fund 'to help make a Better Bristol'. The scheme offered grants of up to £5000 but applicants had to first raise half of their fundraising target via 'Crowdfunder'. This would be a race 'until the money runs out!' Though project workers could receive training from Crowdfunder to launch their idea, there did not seem to be an awareness that this process could never be fair, because wealthier people could email or Facebook all their wealthy friends to help them with the Crowdfund and were, therefore, bound to win the race. It is not surprising, then, that some of the funding went to middle-class groups while excellent projects that were designed by working-class people for working-class people went unfunded, including a plan to make the main shopping centre in Knowle West (Filwood Broadway) more attractive with plants and colour (Bristol Green Capital Better Bristol 2017). Competitions for funding are probably never going to be equitable for working-class people, yet they are still utilised to distribute environmental funds.

Lack of Participation in Decision-Making

An underlying cause of all the problems above is the exclusion of working-class people from environmental decision-making. Environmental inclusion, if done well, could reduce not only environmental inequalities and inequities but environmental problems, more generally. Feminist 'Standpoint Theory' (Smith 1987) made explicit that marginalised groups are socially situated in ways that make it more possible for them to be aware of issues than it is for the non-marginalised. Working-class people have important knowledge and perspectives to offer that can help with addressing wider ecological problems.

The right to inclusion in environmental decision-making is laid out in legislation at a number of levels, in the UK and internationally. 'Environmental democracy' is a strand of Principle 10 of the Rio Declaration (UNCED 1992) and participation in environmental decision-making is the basis of the UNECE (United Nations Economic Commission for Europe) Aarhus Convention which the UK has ratified. Both require access to information, public participation in decision-making and access to legal justice in environmental matters. However, as this chapter has illustrated, there is sometimes a lack of genuine inclusion for working-class people.

It is useful to consider what genuine inclusion would look like. In my earlier work (Bell 2014), I developed a method for assessing the degree to which participation in environmental decision-making is meaningful (see the resulting list of indicators in Appendix A) and in upcoming work I have collaborated to develop a new model of participatory environmental decision-making (Bell and Reed 2019). Inclusion in environmental decision-making, if meeting all the criteria described in the checklist of meaningful participation in Appendix A, would mean that working-class people would be engaged in open and inclusive decision-making process that would allow those most at risk to be involved in scrutinising scientific information, setting standards that better reflect their needs and deciding on the policies, programmes, projects and other interventions that will address these problems. The checklist drew particularly on the work of Sherry Arnstein (1969) who advised as to how to ensure that marginalised groups have control over the agenda and the solutions. Her work describes a 'ladder of participation' with the bottom rungs aiming to 'educate' or 'cure' participants; the middle rungs of 'tokenism' allowing communication but not influence; to the higher rungs indicating increasing degrees of decision-making power, up to full managerial control.

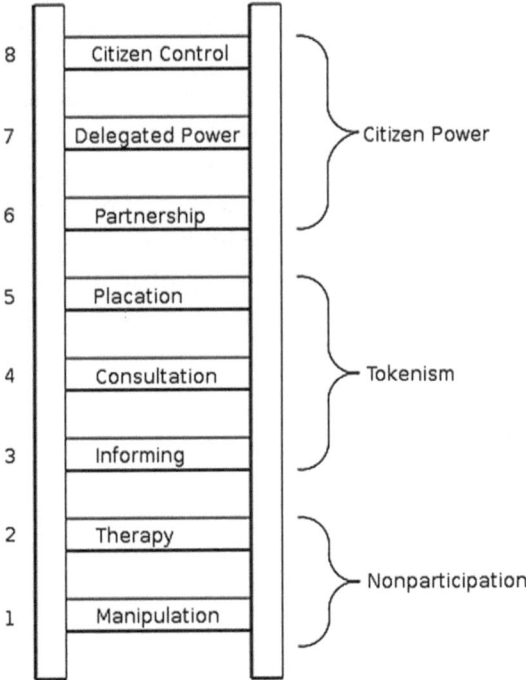

Fig. 4.1 Ladder of citizen participation. (Source: Arnstein 1969)

Arnstein highlights the possibility of manipulation which occurs when the public are 'educated', and participation is in fact a public relations vehicle for power-holders. She critiques the ubiquitous 'consultation' form of engagement which offers no assurance that citizen concerns and ideas will be taken into account. Consultation focuses on how many people come to meetings or answer a questionnaire, rather than the degree to which the popular will is enacted. At the level of 'placation' citizens begin to have some degree of influence but it is quite tokenistic. This occurs, for example, when there are a few hand-picked 'worthy' residents on local partnerships. They are generally unaccountable to a constituency in the community—there is no feedback mechanism from the community to allow proper representation. These token residents are also usually outnumbered by power-holders and the power-holders maintain a right to judge the legitimacy or feasibility of the advice given by these token residents. Delegated power occurs when citizens achieve dominant decision-making authority and citizen control when participants or residents can be in full charge of policy, programmes and management.

We can see examples of all of these forms of exclusion in the above stories articulated by working-class people and, consequently, policies are designed that are inappropriate or, sometimes even, harmful to them. Some of the power imbalance evident here could be offset by strong campaigning and solidarity but, as we see in the next chapter, working-class people are often inadequately supported by environmental organisations.

5

The Environmental Policy Influencers

Previous chapters describe how working-class people are disproportionately carrying environmental burdens and are also often additionally burdened by environmental policy. Environmental campaigning organisations, social movements and green political parties could ally with working-class people to prevent this, but the evidence presented here suggests that they often do not and sometimes their actions compound the problems that working-class people face.

Environmental Organisations and Social Movements

Environmental organisations and movements come in many shapes and sizes and include a vast array of perspectives, from eco-socialism and eco-feminism to nature conservation and green capitalism. Some have a huge amount of financial resources and others have none. Some focus very locally and some internationally. Much scholarly literature is dedicated to classifying the multiple and oftentimes conflicting goals, tactics and organisational structures of environmentalism's many collective actors.

© The Author(s) 2020
K. Bell, *Working-Class Environmentalism*,
https://doi.org/10.1007/978-3-030-29519-6_5

One very useful analysis, by Martinez-Alier (2002b), proposed three main themes as dominating contemporary environmentalism: the 'cult of the wilderness', the 'gospel of eco-efficiency' and the 'environmentalism of the poor'. The first is made up mostly of conservation movements established in the Global North that focus on the preservation of unspoiled wilderness and the restoration of degraded areas. The second, 'eco-efficiency', is concerned with a rational and efficient use of natural resources, using key terms such as 'sustainable development', 'ecological modernisation' and 'environmental services'. The third, the 'environmentalism of the poor' perspective, comes from environmental movements in poor countries or communities, with the environment viewed as material sustenance (further discussed in Chap. 6). This chapter focuses on the classism of the mainstream environmental movement (i.e. the first two categories)—generally funded and well organised and fitting with the dominant beliefs about the causes of and solutions to the environmental crises.

A number of studies show that the mainstream environmental agencies in the UK and US fail to engage Black, Asian and Minority Ethnic (BAME) people and low-income communities (e.g. Burningham and Thrush 2001; Capacity Global 2009). The available literature, my own experience and my research with working-class people in Bristol and beyond indicate two main reasons for this: (1) mainstream environmental organisations tend to be dominated by middle-class people who are unaware of class issues and (2) mainstream environmental agendas can tend to alienate working-class people.

Perceptions that Environmental Organisations are Dominated by Middle-Class People

Marxists would argue that one effect of class domination is the emergence of a distinctive proletarian 'standpoint', or understanding on life. From this perspective, the working-class are considered to be capable of revealing the contradictions in the dominant modes of organising society (Cockburn 2015). As mentioned in Chap. 4, Dorothy Smith (e.g. Smith

1987) and other socialist feminists took up 'Standpoint Theory' as a way of considering women's oppression. They drew attention to the differences in men's and women's lives and highlighted that diverse views from below, rooted in life experiences, were valid, reliable and possibly more realistic accounts of the world.

With this theory in mind, it is significant that the working-class people I spoke to almost all felt that environmental organisations are populated and dominated by middle-class people that would not understand their lives. Mel, one of the women I interviewed, for example, said that she was not at all attracted to environmental organisations, because:

They seem like they're run by very privileged individuals who I would find it difficult to relate to or to communicate with, who perhaps wouldn't appreciate the other challenges of my life … the programmes and advice are being designed by people who do not understand the reality of people's lives on a day-to-day basis and that's why it doesn't work and it's quite patronising to be told, 'You need to do this, you need to do that,' when the people telling you have no real idea of where you're coming from or the reality of your life. (Mel)

Pete, Anita and Dave, among others, made similar comments, stating,

The [environmental] *organising groups are always 25 to 35-year-old graduates … all posh young people … it did feel like* [they were] *a group of friends working with one another.* (Pete)

I'm not only working-class, I'm also a person of colour, and I had seen Friends of the Earth in the UK as just a very kind of middle-aged, white, middle-class group and I had no problems with what they were doing but I just didn't really see myself reflected in them as an organisation or group and I expected their politics might be quite different as well … At international level, Friends of the Earth were doing some great stuff but I still don't find them necessarily a radical voice here in the UK. (Anita)

If I am being frank, I see a lot of the greens as very—as a very middle-class dominated movement. I don't see it as something that is relating to the day-to-day issues of ordinary working people. You know, from what I've seen, they pose the issue in terms of very moral issues about ordinary working people giving up this, and giving up that, but you don't see much sacrifice on the part of the middle-class individuals that are proposing those solutions. (Dave)

However, the main reason the working-class people I spoke to did not get involved in environmental organisations was because they did not know they existed or knew little about their achievements and successes. For example, Julie and Jo said:

I wouldn't have even thought about joining an environmental organisation!... I have heard of a few, like Greenpeace but that was back in the 1970s not from recently.... Friends of the Earth, I have heard of them, but nothing recent. Not to know what they do or anything. (Julie)

I think people have got this image that all environmentalists are sort of loopy, tree hugging lunatics... I have heard of Greenpeace—it's been around for a long time. I am not really sure because a lot has come to light recently about charities and whether the money that they get actually goes to doing anything or whether it goes in the CEO's pockets. I don't know. I don't know much about them but I am aware of them, yeah. (Jo)

There are bodies of literature on particular environmental organisations which come to similar conclusions regarding their class composition. Take, for example, the Transition Town movement, a global movement across 44 different countries (Transition Network 2017). It is intended to increase local 'resilience' to global threats, in particular, climate change (Hopkins 2011). Though the Transition Town network advocates that 'Inclusion and diversity need to be embedded at the centre of Transition as a defining feature' (Hopkins 2011, p. 94), a number of studies from around the world indicate that it is predominantly made up of middle-class, well-educated and reasonably affluent individuals (e.g. Seyfang 2009; Aiken 2012; Merritt and Stubbs 2012; Stevenson 2012; Feola and Nunes 2013; Grossmann and Creamer 2017).

For example, Grossmann and Creamer's (2017) study of the Tooting Transition Town group in London revealed that the group membership did not represent the diversity of the Tooting population. Though the group had an 'open door' policy, it did not actively set out to recruit a diverse range of participants, listen to and enact their ideas, or consider their differing needs. As the authors noted, an open-door policy is insufficient for achieving diversity because it does not take into account some of the barriers to participation. The researchers considered the reason for this lack of diversity was, in part, this 'passively inclusive' approach (p. 162).

In an attempt to rectify their lack of diversity, the UK Transition Network set up a 'Diversity Project' in 2010 and 2011 to broaden participation 'by working with low-income, faith and black minority ethnic groups' (Smith 2011, p. 102). They did not specifically mention class. Focusing on low income can create a sense of reaching out to working-class people while actually possibly engaging a section of the middle-class who may temporarily have a lower income, including wealthy pensioners, middle-class students and downsizers choosing to work less because they have accumulated wealth and/or can opt back into well-paid work when they choose (as described in Chap. 2). Seyfang (2009), for example, focusing on Transition Norwich found that, despite high levels of education, many members had low incomes. She explained this in terms of their 'post-materialist values'—they had basically chosen a low income through downsizing and reducing their work hours. This is quite different to being working-class and being forced to live on a low income. Middle-class people who choose this path often still have the savings, wealth, contacts or earning potential to step out of the situation if it is too painful. As part of the Diversity Project, the Transition Network surveyed its members and preliminary findings indicated that 86% were educated to post-graduate degree level (Smith 2011, p. 102). This indicates very little inclusion of working-class people. While some working-class people do obtain post-graduate qualifications, it is still quite rare, because of the barriers described in Chap. 2.

Some members of the Transition movement believe that any attempt to target certain sections of the population could be construed as being contrary to their apolitical ethos (Grossmann and Creamer 2017). This approach effectively prevents recognition of class and lack of class diversity which is, by its very nature, political. When these lifestyle movements are overwhelmingly made up of the educated middle-class, the danger is not only that working-class standpoints are overlooked but also that 'green' is translated into, or dismissed as, a 'middle-class concern'.

The literature is, again, reflected in my research interviews and my own experience. The working-class people I spoke to had generally not heard about the Transition movement and, if they had, it was not a movement that attracted them. For example, Pete said about his local Transition Town group,

It's mainly a middle-class hobby thing. … it's like The Good Life all over again. It's like, 90% is self-sufficiency stuff being done again by the same people. (Pete)

When I attended a few Transition Town meetings about a decade ago, I also noted this lack of diversity. My attempt to discuss this in the groups resulted in those present reacting defensively, stating that the activities that they suggest are ones that 'anyone could do', such as growing food in the garden. I tried to explain that many working-class people are exhausted by work so, even when they have gardens to grow food in, they often do not have the time to maintain them. They may also be so stressed by debts, health issues and other pressures that any new activity seems too daunting. They may also not have sufficient income to get started. Mel made this point in our interview, saying,

The environmentalists say to grow your own food. I have a really big garden, that would be great to turn over to fruit, and I've done it before when I've had more money. I've done vegetable gardening. I love it. I really enjoy gardening… but I just can't afford even—you know, you have to buy seeds and tools and bedding. (Mel)

The people at the Transition meetings I attended were very well meaning and dedicated but seemed to lack a structural analysis of environmental issues, believing significant change could happen at the individual level alone. When I asked if they could help us with some of the environmental struggles on the council estate where I was living, for example, regarding campaigning to bury the pylons that pass through the estate emitting electro-magnetic frequencies, they gave advice, such as that we should 'take back the land'. They seemed unable to accept my argument that many environmental problems can be solved only by a change of state or corporate policy. Lifestyle change could not solve the problem of the pylons in our midst which seemed to be making people ill. I also attempted to explain that occupying land on a council estate would not be an attractive proposition for my neighbours as working-class people were not usually willing to get a criminal record for a cause because it would make it very difficult for them to get work again. A middle-class person (perhaps when applying for an environmental job) might be able to argue that they had a criminal record for vandalism because they were liberating a field, but this argument would probably not be so readily believed if it were a working-class person. They might simply be seen as dangerous or too rebellious for the kind of work they were applying for. Vosper (2016), in an article entitled 'What Makes Me Tired When Organising with Middle-class Comrades', has written about how, when she is involved in activism with middle-class people, they are able to draw on their privilege when their activism involves breaking the law, for example, in terms of getting character references from people they know in similar positions of privilege, or simply having the finances to pay for solicitors.

Extinction Rebellion has faced critiques that the leaders and activists are 'middle class, self-indulgent people who want to tell us how to live' (e.g. Sky News, 17th April, 2019). Such attacks may, undoubtedly, be used to undermine their message and these activists deserve respect for their concern and their courage. However, they may need to consider the inclusiveness of some of their communication and tactics so as to avoid such criticism. The Extinction Rebellion strategy uses non-violent resistance and 'creative resistance' with some activists pledging to break the law and get arrested if necessary. In recent protests in London, over 1000 activists were arrested (Dearden 2019). Although there must be a place for disobedience in any campaign for social change, the way some of the activists spoke about this disruptive activity seemed to imply that anyone could do the same. When challenged by presenter Nick Robinson on Radio 4's *Today* programme that some people were having a 'miserable' time as a result of the disruption, Gail Bradbrook, one of the organisers, responded: 'they should take some time off work and come and join us' (BBC Radio 4, 18th April, 2019). As well as seemingly failing to realise the limited autonomy to take time off for many working-class people, this comment seemed to ignore a range of genuine working-class concerns, from getting to work on time to avoiding being arrested, as discussed in Chap. 5. As McGarvey (2017) describes, for people from working-class backgrounds there is no margin for error.

Linked to this appears to be a common assumption among environmentalists, and middle-class people in general, that their experiences are universal. They often seem to be unaware of their privilege. One aspect of this is that they don't realise the extent to which cost can be a major barrier to working-class inclusion. Even when the environmental organisation purports to be focusing on equity and equality issues, their activities and events can still be unaffordable. The Schumacher Institute, for example, has always charged very high prices for its public events. Thanks to the work of the class-aware people within the organisation, it has improved vastly in recent years in terms of taking into account social justice issues and now describes itself as 'working towards a world of reducing inequality that provides for everyone within the limits of the planet'. However, the costs to attend its events continue to be too high for many working-class people. For example, as I write, there is a Schumacher event entitled 'All Systems Go', held on 6 September 2017, charging £32.45 for entry, or £21.83 for a concessionary place (Schumacher 2017). Although people may attend without cost, as 'volunteer helpers', this does not necessarily help them to feel equal or give them the freedom to contribute freely. If they wish to criticise or disagree with the main speakers or the dominant view of the discussions, for example, they may feel awkward and disloyal. Even concessionary rates may create awkwardness about receiving 'charity' and can be perceived as patronising. I have often observed that middle-class people seem proud of their ability to get something for nothing (food from skips, for example), while working-class people are proud of their ability to pay. Putting on affordable activities is basic to good equalities practice with working-class people.

In February 2019, we saw the first of a series of school strikes to protest against climate change, inspired by Swedish student Greta Thunberg. School students in 40 cities across the UK joined those in the US, Australia, other parts of Europe and Africa to call for more government action. In the UK, the actions were organised by a range of young people's organisations, including YouthStrike4Climate and the UK Youth

Climate Coalition. Although there has been no class-based analysis of the movement as yet, it may have been difficult for some working-class children to attend for a number of economic reasons, including that working-class parents may be more afraid of being fined for their child not attending school; they may be less able to take time off from work to transport their children to the event; and they may not be able to pay their children's transport fares to the event. While some parents proclaimed their willingness to pay any fines for their children protesting (Morss 2019), others would obviously not be able to do this. Being active on these issues at such a young age is really important for involvement in environmental campaigning throughout life and so it would be important to consider how to include working-class young people in the protests.

If working-class people do overcome the financial barriers, a further difficultly they face is that, as Vosper (2016, np) notes, 'Middle-class people can tend to dominate meetings, especially at public events. There is a sense of entitlement that the whole world needs to hear your opinion and you have all the answers'. I have also noted this tendency in many environmental groups. For example, in the Green Capital Inclusion Group, introduced in the previous chapter, there were three particular White, middle-aged, middle-class males who dominated the meetings each month. Not taking into account the fact that it was a group about inclusion, they spoke in relation to almost every agenda item, even though the agenda was always full and the meetings large, with many people waiting to speak. On one occasion, when I was asked to chair the meeting, I moved on to the next agenda item even though one of these males had raised his hand to speak about the current item. On this occasion we were running late so to move on seemed a legitimate decision. However, he immediately interrupted me, complained about unfair treatment in not having the space to speak whenever he wanted and stopped me moving on, making his point despite my decision.

Of course, not all middle-class people behave like that and some in the Green Capital Inclusion Group did listen well, including the men. I am discussing a tendency to dominate rather than behaviour that applies to all middle-class people. This domination can even, at times, be aggressive, or at least perceived as such by working-class people. All the working-class people that I interviewed who had been involved in environmental organisations reported at least one experience of being 'told off' by middle-class people in the group. For example, Mick complained that 'They stop you half way through your sentence and say you are not using the right word' (Mick). Working-class people often mention having their grammar or other aspects of their language 'corrected' by middle-class people as an aspect of classism. Usually the working-class person knows that their grammar is not technically correct, but it is a form of cultural expression, so such comments effectively impose a middle-class language code on the speaker.

The correcting can at times be carried out in a way that comes across as patronising and morally superior. For example, Pete described an incident where,

> *There's a person I thought was a woman and* [I did not address them correctly so] ... *the person said, 'I'm gender neutral, and my pronoun is V', or something* ... *I apologised and after* [I went to speak to one of the leaders] ... *and they went, "Well, just* think!*" like a Sergeant Major ... I've still got things to learn, but the policing of that ... I was being told off.* (Pete)

Linked to this issue of moral superiority, many of the working-class people I spoke to said they, or others they know, have become suspicious of, or even hostile to, environmentalists because they perceived them to be 'preaching'. For example, Mick said:

> *We knocked on doors to tell people about the generators, but there was massive suspicion and rejection of anyone who looked 'green'* ... *When they went to working-class areas they were not greeted, they were confronted* ... *Green is a problem* [for working-class people] *because people feel they're being told what to do but they are already doing as much as they can. We already use public transport, but the public transport doesn't go to where we work, it just goes in and out of the city centre. We drive taxis for a living and, because it's diesel, we get a lot of hassle for that. We cycle, but in a working-class way, on an old treader, no racers, no lycra* ... *The long and the short of it is that all the standard green solutions are either making our problems worse or are irrelevant. Muslim women with long clothes and a large family can't do it on a bike.* (Mick)

Similarly, Sharon said:

... there's no give sometimes 'cos this person that I'm thinking of [an environmentalist] *used to say, "You should all be doing this and you should all be doing that" ... That's not good. That's not involving people ... They're not working-class and so it's a different world and the way that they speak is so academic and it's not relevant, I don't think, to a lot of people.* (Sharon)

Mick also alluded to this air of moral superiority that he perceived middle-class environmentalists to be projecting:

They think some are more entitled than others ... [That is] the traditional, white, middle-class environmentalist who believe in meritocracy, entitlement... We just see them coming out of work in their lycra, getting on their bicycles, grumbling about all the pollution from cars, coming home, ignoring their taxi driving neighbour who might be out washing his car, about to go on the night shift and deal with all the drunks, and then put the wood on their wood burning stoves. (Mick)

By mentioning the polluting wood-burning stoves (see Chap. 4), Mick implies that their perceived superiority is not justified and is, actually, hypocritical. He portrays them as rather self-centred and dis-interested in working-class concerns.

Many of the working-class people I spoke to also conveyed the feeling that, while the middle-class environmentalists seemed to criticise them for their lifestyles, jobs and culture, they could also appropriate that culture, without recognising where it had come from, as Sharon commented:

The working-class could teach the middle-class a lot about how to live without waste, when you're skint, but what pisses me off is that, you know, you have to turn organic and artisan ... and now the middle-classes have come along charging £20 for a loaf of bread that only they can afford ... They've even taken away, do you know what I mean, all the things that the working-class used to do, like recycling clothes... the working-classes had all those skills and now they've all been taken away and they're all being charged loads of money for it ... I'm just like, well really? ... So, they hate our culture, but they take it away from us [for themselves]. (Sharon)

In general, my research respondents said that middle-class people tend to look upon the working-class as needing to be helped, rather than recognising their skills, and treating them as equal:

We had a discussion about which skills, what kind of skills people had and, interestingly, the [middle-class] *people said that people off the estates had some of the skills that they actually needed like first aid certificates, food hygiene certificates, do you know what I mean? So, people could actually help each other but the assumption before that was that the people off the estates will need help.* (Sharon)

When they do reach out [the environmentalists], *it's kind of like, okay, those White guys are coming to save the poor Black people, you know. It's kind of this patronising expert knows best, you know. … I'm not interested in somebody coming along and telling me how to behave and what to do and telling me I'm not doing enough.* (Mel)

The general thing about the organisers of this [environmental] *movement, they're not thinking about the working-class. They are not working with them, they're not thinking about how to mobilise them, they're not working with them as equals… For them* [equality] *means not using a word, it doesn't mean listening to people. They think they have all the answers …. "I know best". "We're coming in to help you".* (Pete)

The point that mainstream environmentalists are not organising with working-class people on an equal basis resonates with my own experience of the environmental movement. I will give an example here in relation to the Green Capital Inclusion group, discussed in Chap. 4. As I described, an 'Inclusion Group' was set up for the Green Capital programme in the early stages. I attended the 'Inclusion Group' as a representative of my local environment group upon our own initiative—our group was not invited even though we were, and still are, the only general environmental group on the estate. My intention was to advocate that we use the Green Capital as an opportunity to rectify some of the environmental inequities across the city (e.g. those discussed in Chap. 3). At the inaugural meeting of the Inclusion Group, it seemed the participants were mainly representatives of local environmental NGOs, as well as some city councillors and council officers. I heard mostly middle-class accents and there were very few equalities leaders or people from the deprived inner-city areas or the council estates present. I do not wish to make invisible the working-class people that did attend with this comment and, obviously, I could not say for certain the class backgrounds of the participants, but this was my perception.

The attendees of the first meeting, with a few exceptions, expressed their purpose for being there in terms of 'getting lots of people interested in Green Capital' (comment from first Green Capital Inclusion Group attendee, 2014). A number of events intending to change individual life-styles towards being more sustainable were suggested and considered at these early meetings (including, for example, a picnic on the M32 motor-way). These suggestions, and the associated discussions, indicated little understanding of any equalities aspect to the terms 'inclusion' or 'green' on the part of the majority of the group, including, and especially, in rela-tion to class.

When I endeavoured, by making suggestions at the first three meet-ings, to ensure that Bristol's equalities groups were involved (including representatives from the deprived inner-city areas and the council estates), there was resistance from the majority, but not all, of these environmen-tally motivated attendees. I was told initially that it was 'not necessary' or that they had 'already tried' and later that it was 'too late' (second meeting of Green Capital Inclusion Group) to involve these groups. Throughout the meetings, when I continued to speak on the opportunities implicit in Green Capital for tackling the environmental injustices faced by these equalities groups in Bristol, including for working-class people, dismis-sive comments were made and I was told that we should focus on 'cele-brating the resilience' of these groups, rather than pointing out injustices (third Green Capital Inclusion Group). One of the few working-class people who had attended from the beginning dropped out because she found it too stressful to be in the group, in part because she felt that the working-class people who had come were not treated with respect.

However, I continued to advocate for more inclusion and was finally given 'permission' to contact the excluded groups until, slowly, the group became more diverse. Eventually I also left, though, because I felt that the group was still too dominated by the ideas and desires of the middle-class environmentalists who wished the Green Capital event to be a 'celebration of all the diverse environmental projects in Bristol', rather than tackle some of the dreadful environmental problems that impacted on the working-class areas.

Mostly, it seems, the working-class experience of environmentalism is one of having to find the organisation themselves, make whatever sacrifices are necessary to get involved with them and then 'fit in', as Pete and Amy described:

> When I try and get [middle-class] environmentalists to go out and talk to people, they sit down and go, "Why don't people come to our meetings?" No, you have to go to their meetings, and you have to ask them what their concerns are, and work from there, and it just doesn't seem to be—it's not an attitude which the middle-class people who run these Friends of the Earth groups, or whatever, have … It's just a big struggle to try and bring these topics up … You bring it up and … they don't want to know, that's how it feels to me … they're not trained to do it; they're not thinking about it … it just feels _draining_ … I give up. So, that's where I am with that one. (Pete)

> They [environmental organisations] don't have any tangible presence in this town at all … there is nobody here representing them. I like anything that gives people information. If they set up a stall really central to the town centre, I think people would show interest and curiosity … but, no, I have never seen them here. (Amy)

The lack of internal democracy within some of these organisations can further undermine a working-class person's chance of being heard. One of the ways that middle-class environmentalists tend to organise that can be problematic for working-class people is to try to work without any formal structure. A study of Greenpeace, for example, showed that, for the first ten years, decisions were made on an ad hoc basis. This lack of structure, unfortunately, 'created the basis for a fundamentally undemocratic organization in which decisions were made by a small group of people, predominantly men from the professional managerial class' (Harter 2004, p. 90).

The middle-class environmental organisations, while being loose on structure, are fairly rigid when it comes to emotional expression, however. Nicole Vosper observes 'this inherent need for middle-class people to censor, control and mediate emotions' in the activist groups she works with (Vosper 2016). Lakey (2010), an activist trainer as well as an academic, argues that middle-class control of expression unfolds in groups time and again. He considers this is because economies operating on class lines require a great deal of sorting, screening, correcting and grading of people. In class societies the middle-class must manage the workers and so, in the groups he runs, the person most likely to take on this role of controlling emotions are significantly more likely to be from middle- and owning-class families. If a working-class person takes on this role it is because they have graduated from college and assumed the values of management and control. They have taken on the supervisor's duty of correction. Lakey (2010), therefore, notes that those who most rigidly hold back their feelings and have fully mastered self-control are those whose position puts them in a dominant position to others—White people, men, owning and middle-class and heterosexual people. Those who enforce suppression of emotion are supposedly teaching us to be appropriate. The problem is this stifles authenticity. Pete also spoke about this problem, saying that, in the environmental groups he had been involved with,

There was things about not being angry ... I've heard that more than once, if you're angry, what you should do is do what good managers do and have a word with the person and take them quietly to one side. They never have an argument in the open, if that makes sense? ... There is an absolute inability to deal with anger. (Pete)

Controlling harmful expressions of anger, such as shouting and name-calling, can help people to feel safe in a group, so it is useful to do this. However, Pete and several of the other working-class people that I spoke to felt that the middle-class people in the environmental groups they had engaged with often expressed anger in more harmful ways that were difficult to challenge. Rather than directly saying how they feel, they use passive-aggressive methods such as the 'silent treatment', correcting, or continually ignoring and rejecting an individual's ideas and requests.

All the above reasons explain why predominantly middle-class organisations could have a problem engaging working-class people or recruiting them and keeping them as members. Middle-class environmentalists could be admired for caring about the planet, wanting to help people and being prepared to make sacrifices for their beliefs, but they are, instead, perceived as being unwilling or unable to engage with working-class communities on an equal basis so that their environmental message goes unheard along with their potential for inspiring others. This is obviously problematic in terms of building a wider environmental movement. Some organisations within the environmental movement now recognise this and aim to do something about it. For example, Craig Bennett, CEO of Friends of the Earth since 2015, said in an interview

> *I feel passionately that, to up our game, the environmental movement needs to make sure it is not just a white middle-class movement. We need to make a really big effort to look at how we can reach out to different communities and create opportunities for them to campaign on the issues that matter to them.* (Bennett 2015)

However, as yet, mainstream environmental organisations, in general, have been slow to acknowledge and act on this problem. The reasons for this and the possible solutions are discussed in Chaps. 7 and 8.

The second main reason that working-class people tend not to engage in the mainstream environmental movement is the focus of the mainstream agenda which can be off-putting for some.

Environmental Agendas

The class composition of environmental organisations affects their campaigns. For example, a study by Harter (2004), found that the social class of Greenpeace members affected their politics and choice of activities. The founding members of Greenpeace, and those who would become the most well-known initially, all worked in middle-class professions and this class composition changed little over the years. Harter noted: 'Despite their claims to be democratic, universal, and above class interests, the methods they choose often have very real negative effects on one particular group: the working-class' (2004, p. 92). He illustrates this by describing how their campaigns on hunting and logging focused not solely on the large profit-making companies but also on the activities of working-class communities struggling for survival. In the late 1990s, for example, Greenpeace came into conflict with logging unions by blockading them from carrying out their work, thereby preventing them from earning a living. The loggers had already been badly impacted by technologies which brought in faster, more profitable and less environmentally sound logging. These technologies had eliminated jobs so that, while production continued to grow, employment decreased. The logging companies encouraged workers to believe that the environmentalists were to blame for the job losses so that workers and environmentalists were played off against each other. Unfortunately, this could happen because Greenpeace did not include the actual communities affected in developing their strategy.

Similarly, other environmental groups in the US have come into conflict with loggers, reportedly telling them to go and work at Walmart (Loomis 2016) in an evident disregard for the pride and identity that some working-class people feel about their jobs.

On the whole, it might be expected that, due to the pressure of 'cognitive dissonance, people who are employed in environmentally damaging industries will justify it to themselves. A comment made by one of the interviewees, Jack, who works in aircraft production, illustrates this:

> *If I could get another job that was of similar interest, that was environmentally, you know, correct, then I might think about it ... I mean, it would have to be a job that used my skills and that I could get enough money for ... The thing is, if I gave up my job, it's not going to stop them making planes ... People need to travel in the modern world. We should just try to make planes that are better for the environment and they are trying to do that.* (Jack)

Jack was prepared to do other work but he did not want to lose his skills. In other parts of the interview it was evident that he identified with his work and felt proud of what he did for a living. He was well paid but it was not just the pay that motivated him; it was feeling part of the workforce and feeling valued at work. Joe Uehlein of the US Labor Network for Sustainability asserts that environmentalists 'have never been able to understand the primacy of work in people's lives so their messaging is tone deaf to the needs and aspirations of working people' (Labor Network for Sustainability 2018). Sometimes there are no other jobs in a community except in the environmentally harmful industry, so to critique these jobs without discussing alternative possibilities with the workers themselves is very likely to alienate them. It has been reported that, in one incident, Earth First! drove metal spikes into trees, nearly killing one millworker, which only reinforced the opinion among the loggers that environmentalists did not care about them (Loomis 2016). As Räthzel and Uzzell (2012) point out, work for many not only is about survival but also provides dignity, identity and solidarity so that when industries are attacked as environmentally harmful, those who work in those industries will also feel attacked (Räthzel and Uzzell 2012, p. 84). While the US environmentalists were focused on the workers' misdemeanours, they were also ignoring the workers' problems, such as automation and outsourcing, both of which have decimated the US working-class (Loomis 2016, np).

Environmental organisations have occasionally aligned themselves with trade unionists and workers in accord with the workers' own agenda. For example, Greenpeace supported the Bhopal gas victims by sending their ship, the *Rainbow Warrior*, to Mumbai to increase global awareness of the disaster and supporting the locally based Movement for Justice in Bhopal (MJB) with research and publicity. The MJB were campaigning for justice following a toxic gas leak from a pesticides factory operated by an Indian subsidiary of the Union Carbide Corporation (UCC), which led to 7000–10,000 deaths immediately and a further 15,000 deaths over the next 20 years. The local organisation engaged in a range of campaigns from legal actions to hunger strikes and marches. However, even in this positive alignment, there were complaints that Greenpeace took credit for the local organisation's activities and tended to dominate the public debate (Mac Sheoin 2012).

In general, the mainstream environmental organisations commonly tend to be focused on their own agenda, without thinking how it could mesh with that of working-class people. Very often environmental organisations were perceived by the working-class people I spoke to as yet another demand with their requests for money or time, when people were already feeling overextended. They rarely seemed to support the social or environmental initiatives of working-class communities. Mick, who lives in an inner-city area, one of the poorest places in the UK, was facing a problem where it was proposed to put polluting diesel generators in their locality. Mick tried to get mainstream environmental organisations involved with a local campaign to resist the siting of the diesel generators there with little success. He said:

> *When they wanted to put 24 diesel generators in* [local area], *we set up an organisation … with the people living downwind from the generators …* [but] *… Friends of the Earth, the Green Party, all the great and the good in the environmental movement, we had a massive problem getting them involved. Just a few saints among them helped us.* (Mick)

Mainstream environmentalists often emphasise the need to modify our individual behaviour, especially in relation to climate change. There is a focus on what we need to give up and stop doing. The rationale for this makes sense to the extent that, as Chap. 1 highlighted, globally, we are consuming resources and emitting waste at a rate that will lead us to irreversibly overstep planetary boundaries (see also Meadows et al. 1972, 2004; Latouche 2009; Jackson, 2009). A recent study (Raftery et al. 2017) estimates that, though many countries are now decarbonising their economies, the chance of keeping global warming to less than 1.5 °C is just 1% because of global per capita growth rates. Almost all the technical improvements and efficiency will be cancelled out by this rate of growth. The amount that we need to cut back is substantial. Fisher (2013) estimated that affluent countries need to reduce their energy and material consumption to 6% of what it is now for humans to have a habitable planet in the long term. We must face up to the fact that high-consumption lifestyles can never be available to all, even if we had 100% renewable sources of energy. Some environmentalists, therefore, advocate that we must therefore build a 'simple-lifestyle' campaign to create a 'readiness to sacrifice luxury and comfort' (Sarkar 1999, p. 229).

Individuals are encouraged either to engage in 'green consumption', or 'voluntary simplicity', that is, to consume differently (e.g. different light bulbs, food or travel), or to stop consuming as much in general and live more simply. A whole range of lifestyle and grassroots responses to the environmental crises have, therefore, emerged in recent decades including Slow Food, Permaculture, eco-villages, Transition Towns, community gardens, eco-building, 'farmers' markets, local currencies, co-housing projects and skill-sharing networks. Many of those involved in these projects tend to believe that, in addition to building resilience and avoiding environmental harm now, they can help to generate the values and ideas that can feed into longer term political reform. Therefore, they consider these projects to be not only lifestyle approaches but also activist responses.

However, when environmentalists talk about reductions in income and material wealth and even a shorter working week, which many equate with a loss of income, this is not a positive message for working-class people who, as described in Chap. 2, often experience a lack of economic security. This is not to say that an orderly economic contraction or degrowth is not desirable or achievable (see Chap. 8) but it has to be discussed and achieved in a meaningful way which delivers benefits for working-class people at the same time, and not later. At the moment, environmentalists are advocating a reduction in resource and energy consumption while markets continue to determine the allocation of resources. Such a contraction would almost certainly lead to inadequate profits and, consequently, companies going out of business and increasing unemployment levels. Therefore, working-class people rightly resist and question the idea of reducing consumption because it currently equates to losing jobs. The last chapter of this book will discuss how degrowth or postgrowth can be good for working-class people, through linking it to much wider structural change that will bring about greater equality.

In the meantime, these lifestyle approaches can, and often do, alienate working-class people. They tend to imply that everyone has the same ability to carry out these virtuous acts, even though many working-class people do not have sufficient income to engage in, often expensive, green consumerism or sufficient time to take part in time-consuming acts of 'voluntary simplicity', such as growing their own food as discussed earlier. They may also lack sufficient health to make environmental choices, such as cycling and walking to work or may, simply, not feel safe enough to make these choices. The working-class people I spoke to considered that many aspects of their lives are negatively judged by environmentalists, from the food they eat, to the way they travel, to the work they do. If working-class people try to explain to middle-class environmentalists their genuine reasons for not being able to be green in certain circumstances, their comments can be dismissed in ways that reflect a lack of understanding and sometimes, even, a lack of basic respect. For example, Mick, Mel and Jo remarked:

It's green to force people out of their cars… therefore irrelevant to address the fact that a woman has to drive her kids to … avoid exposing them to anti-social behaviour… being spat on because you wear a Hijab. (Mick)

I think it [consuming less] *is massively important… The amount of clothes that we go through, the amount of landfill that creates, the amount of energy that's used to create clothing… But I still regularly buy at Primark because I can't afford the higher quality clothes. So, I totally agree that we need to consume less and that it has a cost, but with most of these campaigns … the people who say this, they are not living on a budget… We're not talking the same language … You go to* [cheap clothes shop], *and spend three pounds on a pair of shoes that, yeah, will break in a week or two weeks' time, but by then you might have another three pounds to spend. In the long run, it's more expensive, but that's the reality.* (Mel)

Free range eggs and stuff like that are more expensive but I buy them because of the cruelty [of the battery system]. *I don't buy free range chickens because they are quite expensive. I should use more eco-friendly cleaning products but I don't because I shop at Lidl and I buy the cheapest ones that I can find … I think if we had more money it would be easier to buy the more eco-friendly stuff, definitely, but I can't so we do what we can.* (Jo)

Anita, who as described earlier, has now acquired some traditionally middle-class tastes, discussed judgemental attitudes to working-class people who consume more when their economic situation improves:

They mostly lived in social housing. They've now bought properties. They can't understand why I would live with a bunch of other people communally in a house. They can't understand why I would ride a bicycle. Now, that's not that

... they don't care about the planet, but also there's something about how the system is set up that we aspire to certain things and so, as we might improve our social economic conditions, there are things that we buy or we want that are actually quite harmful to the planet. But middle-class people get to have them and not be judged. (Anita)

Middle-class environmentalism has also been accused of fetishising poverty or talking about it as a game or adventure. As Vosper (2016) states, this 'is an insult to folk who have no choice ... unlike middle-class people we don't have a safety net. We can't play the romantic poor anarchist for ten years then inherit property. Flirting with poverty as a lifestyle choice is not the same as growing up in poverty' (Vosper 2016, np). In particular, calls to live without money can be very offensive. For example, in Bristol, we had the famous 'Moneyless Man' who wrote a book about how he lived for a year or so without money, advocating it for others (Boyle 2010). Some of the middle-class people I have met, while having a good income, are proud of their ability to source free foods from skips. It is good that the food does not get wasted, but they often present it in terms of the money they have saved. Since they do not need to live on a low income or without money, it seems as if they are pretending to share an experience of poverty which is not authentic. Working-class people, on the other hand, often do all they can to hide their poverty. McGarvey (2017, p. 23) notes when writing about his childhood in poverty, 'the deep sense of shame many of us felt about our poverty' and how buying all the goods to appear better off was expensive 'but the price of looking poor was higher'. Being genuinely poor is incredibly stressful and can kill—it is not a fun thing to do for a while. The Moneyless Man, and similar others, fail to take into account all the support and help they get from their middle-class contacts as well as free education and healthcare paid for by taxpayers.

The individualistic focus of much of environmentalism, evident in these calls to make changes to our lifestyles, can also alienate those working-class people who see the problems as more structural. For example, Anita and Dave explained,

... the environmental issues that we have aren't simply about let's reuse—let's not have single use plastics and things like that. Obviously, that's all very important

but I see environmental degradation as very much more systemic and as a kind of one of the many bad bi-products of capitalism … I think when we look at environmentalism, the language of it, that's very much down to lifestyle rather than systemic. We look at it systemically, I think working class people have more to gain from systemic changes and middle-class people have less to gain. (Anita)

I see the environmental issue as in a context of a whole number of other issues, and that is why I'm a socialist, and I believe that the only solution is getting rid of capitalism. So, I'm not in tune with the Greens or people who see environmental issues as a single issue, because I think that those issues will not be resolved until you actually remove the profit system itself … I think there's some very good people within the Greens, some very sincere people, but where I draw a dividing line is when you then start to separate that issue off from the broader issue … I cannot see a solution to the problem, as long as the existing system goes on, because the principle that dominates everything else is profit, and that is how I would see my difference with a lot of people that would term themselves environmentalists—they do see a solution in terms of the present system, and I don't. (Dave)

As set out at the beginning of this chapter, there are as many streams of environmentalism as there are, literally, shades of green. However, generally speaking, as Anita and Dave's statements describe, the mainstream environmental movement (rather than 'the environmentalism of the poor') does tend to advocate solutions that do not involve radical structural challenges to society.

Overall, then, my research and experience show that environmental organisations are not generally attractive to working-class people because they do not work in ways, or work on issues, that working-class people feel comfortable with or can relate to. The next section discusses the Green Party, as a specific environmental organisation, where similar problems can be found.

The Green Party

Here I will focus on the Green Party of England and Wales. I am reluctant to criticise the Green Party because I believe it currently has the best environmental policies of all the parties in the UK and many excellent social policies too. It also has many very well-meaning members, some very dedicated and inspirational. However, generally speaking, to date, it has not been immensely popular in the UK and some reflection on why this might be seems necessary. Most of this section is about my own experience of being in the Green Party because the working-class people I spoke to in my research mostly had had nothing to do with the Green Party and only one had actually joined. Not all experiences were negative. Pete and Amy, for example, though never members, had quite positive impressions of the local or national Green Party, saying

> I think here, the Green Party—the keen ones, the very keen organisers will go out and work with working-class communities. They will support their campaigns. They will go to any meetings … It's a mixed picture, but I know they get involved with local campaigns and they do go to community groups. They will work with trade unions. (Pete)

> I have voted for the Green party. I like the fact that they're trying to do something to take care of the environment and they're giving it a better emphasis. People just wouldn't pay attention if there weren't people out there speaking about it. (Amy)

However, as with the other environmental organisations, the most common experience was that of a lack of information or presence in relation to the Green Party, as Julie states here:

> [The Green Party]—I don't know what they do really. It is just another party. Another political party. I have never spoken to someone from the Green Party. I don't know if we have got anybody round our way in the Green Party. I think we have only got the Labour, Conservative and Liberals… I don't think we have got a Green Party. We might have but I don't remember seeing them on the list. (Julie)

Several of those I interviewed said they had voted for the Green Party, though they had now stopped voting Green because they saw it as a single-issue party or a party that was currently too marginal to get into power, as described here:

I voted for them on one occasion. Only once, because there are so many aspects of what you should be voting for, other than just the green [issues]… It should be proportional representation and then they would have more of a chance. (Bob)

I think, even though I wholly support what they [the Green Party] say in their policies, what is important to me is the other things. I would say my perception, it could be very, very wrong, is that wealthier people, perhaps, may be more able to engage with the Green Party… I think the people on this estate [are concerned with] getting electric, food, bus fare, keeping your internet going so you can put your claim in for your Universal Credit or working four hours a week on a zero hours contract and having to get a bus six miles to do it. I think those are probably more pressing issues … people do not have much money, so I think it would be low on their list of priorities. (Julie)

Although energy, food and transport are clearly environmental issues by most definitions, Julie and others seemed to be perceiving the Green Party policies as being about issues that were more concerned with other species and generations than themselves at this point in time. Perhaps this is a result of the dominant culture that separates humans and nature but also the Green Party could maybe do more to explicitly make these connections in their communications. In other parts of the interview, Julie expressed her deep concern about environmental matters, but she supported Labour because she felt it prioritised the immediate problems of her community. Like many others I interviewed, she was not aware of the social policies of the Green Party that would apply to these problems.

I have also voted Green and, in the past, have been very inspired by the Green Party. However, my experience of being a member of the Green Party was difficult, primarily I believe, because of class issues. I have been a member twice, once in the 1980s and again from 2009 to 2016. For a while I was 'Diversity Champion' for my local branch of the Green Party of England and Wales and it is this experience I will draw on here. As Diversity Champion, it was my role to bring an equalities perspective to the work of my local branch. However, I decided to resign from this post after three months because I could not influence the branch to make the changes I considered necessary. For example, in a policy discussion about recycling, I suggested that green waste collection should be free at the point of use, in the same way that landfill collection is, because many of the people on my estate could not afford the charges (see Chap. 4) and burnt their waste instead, creating health hazards. All of those who spoke in the meeting resisted this idea, seemingly unable to accept that people were unable to afford a service. One middle-class group member responded by stating that the people who live on my estate 'need educating'. They did not believe that people did not use the green waste collection service for economic reasons and not because they lacked knowledge or intelligence.

In a private conversation with one influential member of the local Party, I was told, 'We are already doing a really good job at including diverse groups of people—lots of Somali people voted for us'. I asked him whether he had invited any of them to the meetings to which he replied, 'no, they have different values, look at the way they treat their women'. He was a public-school-educated White male. When questioned, he seemed to have no evidence regarding ill-treatment of 'their women' and so appeared to be dismissing an entire national group on the basis of his own prejudice.

In another example, I said that perhaps we should drop the 'atunement' at the beginning of meetings. This is a period when people are supposed to sit in silence at the commencement of a meeting so they have time to 'tune in'. I explained that working-class people would probably find the atunement unfamiliar and the first thing most people want when they arrive in a new situation, especially if it is outside of their comfort zone, is to feel they can relate to what is going on. In working-class situations, people tend to greet each other with jokes and smiles and, sometimes, physical contact. It would seem rude to sit in silence, apparently ignoring a new person that comes into the room. In response to raising this issue, I was told that 'we must stay close to our core values and not change them just to include new people'. The general attitude was that newcomers must fit in. The majority of the group in the meeting where this was discussed seemed to feel there was no need to change their practices so as to enable a more diverse group of individuals to engage. Like everything else I said about diversity, this point was also not accepted. People became defensive and somewhat aggressive.

In one of the most difficult meetings, when I suggested that the Green Party campaign on the council estate where I live or any of the other council estates on the outskirts of the city, I was told that these places were 'too difficult to get to' for the Green Party campaigners and that we have to target known winnable wards. Council estates were not considered winnable by these Green Party members. When I suggested that we have the meetings closer to the council estates (or even on them), I was told, 'you just want to get the meetings closer to your house'. When I said one way to get a diverse Party is to invite people from different backgrounds to our meetings, I was told it is not a good idea to bring someone from a different background to a meeting because they would 'find it boring'. When I suggested that we should set up a monitoring system to see how diverse our branch and party were, I was told that it would not be a good idea because 'it might upset someone'.

As a result of these and numerous other similar incidents, I became conscious of the extent of environmental classism in the Green Party. Still hopeful that I could change this from the inside, I decided to organise a fringe event at the National Green Party conference on 25 February 2011, entitled 'Working-class Representation'. We asked people: 'Does the Green Party sufficiently represent and include working-class people?' Do working-class people feel at home in the Green Party? What working-class issues should the Green Party focus on? About 20 people attended and almost all the working-class people present (of which there were few) said they had not felt comfortable in the Green Party for different reasons, mostly with regard to it being an alien 'culture', in terms of ways of speaking and taken-for-granted assumptions, and there not being enough people from a similar background as themselves for them to feel comfortable. However, it did seem that some branches were much better than others and one working-class member who is now very high in the ranks of the party said that he felt he could 'get away with a lot because they just put it down to me being working-class'.

As a result of this workshop, I thought perhaps the problems I had encountered in my own branch may not be representative nationally, so I focused again on getting some local-level change. I wrote a report to the branch reminding them of all the incidents described above. When I delivered the report, it did not go down well, although a few people spoke in support of what I was saying. Those present at the meeting said the Green Party had its own unique culture that they did not want to change. One middle-class woman academic said that society is too diverse now for class to have any meaning. Several said class was not an important or relevant issue. After the meeting, in a private conversation witnessed by others, one member accused me of trying to destroy the Party locally.

Later that evening, I wrote up all that had happened and sent a report to the National Diversity Co-ordinator. As a result, some concerned Green Party members working at a national level invited me to speak at a fringe meeting on Class in the autumn conference on 11 September 2011. At this workshop, I told people about my experience in the Green Party and described some of the issues that I have laid out in this book. The room was packed. After I had spoken, a few middle-class people insisted on a debate about definitions, bringing up some of the objections covered in Chap. 2 (hence, I felt the need to write that chapter). However, the working-class people in the room stood up and made very similar points as I have made here, about feeling excluded and uncomfortable in the Green Party. The video of this session was available on YouTube but is no longer to be found to the best of my knowledge. There is only a reference to the fact that the workshop took place (Green Party 2011).

These occurrences are consistent with the literature which describes how those from privileged groups tend to deny the existence of oppression (for a good summary, see Goodman 2011). They often dismiss problems raised by a member of the oppressed group, labelling the person communicating the problem as oversensitive or a 'troublemaker'.

If the members want to build the Green Party, they will need to better engage working-class people, especially now that the Labour Party has become more politically left and is developing better environmental policies under Jeremy Corbyn's leadership, including the 'Green New Deal'. The only working-class person that I spoke to in the research for this book who had been in the Green Party said on this topic:

> Yeah, I was [in the Green Party]. *I met some really wonderful people there, but I was so disappointed that, what they found very difficult is … it's a very, kind of intellectual party, it seemed to me. They had no means to communicate with,*

you know, any minority or even working-class, you know, not-so-articulate people in the working-class … I had an idea, when I joined with the Green Party, to get the message into the working-class areas… only two of the Green Party came to support me, you know what I mean? So I realised that, it was something that they weren't able to do, weren't able to get out on the streets … They weren't prepared to go that extra mile to engage. (Jim)

Like myself, Jim eventually left the Green Party in frustration. A few years after the meeting about working-class inclusion at the National Conference, the Green Party did undertake a survey of its membership which asked questions about class. The results were held up as showing that 'the Greens are becoming the true party of the working-class' (Mortimer 2013). This claim was mainly based on the number of members perceived as living on low incomes, because over 60% of members earned less than the median income (£26,500 at the time). However, the proportion of members with a university degree was found to be 57%; 37% had a master's, PhD or other 'higher' degree; and 56% self-defined as 'lower middle-class'. Of course, middle-class people are much more likely to respond to surveys, so it is difficult to draw conclusions from the survey alone, but the writer evidently had little understanding that middle-class people can temporarily have low incomes when they are students, or pensioners, or 'travelling' or 'downshifting'.

In this account, I have focused only on 'microaggressions' rather than Green Party policy. There is not the space to make a comprehensive assessment of Green Party policies in terms of their benefits or not for the working-class. In general, as a self-proclaimed anti-austerity party, their policies should be beneficial for the working-class. Three out of four of the Green Party councillors voted against the £35 million in cuts to local jobs and services in the 2013 budget in Bristol City Council (Green Party 2013), for example. The Green Party have also long supported benefit recipients, living wage campaigns and the protection of public services.

Furthermore, since I was active in the party, a number of changes have been made which might help to make it a party of and for the working-class. For example, in 2013 the Green Party of England and Wales voted for a statement of core values in the constitution to include a commitment to social justice and a 'transformation of society for the benefit of the many not the few', with 71% of conference delegates voting in support of the change.

It is also important to emphasise that the Green Party is far from homogenous in terms of opinions and approaches. The members that most supported my work on diversity were from the section of the Green Party called 'Green Left'. This section's mission was to be a counterbalance to the lifestyle environmentalists in the party and they include anti-capitalists, eco-socialists and other radicals. Among other pro-working-class activities, Green Left campaigned vigorously against council cuts, especially those implemented in Green Party controlled Brighton. Over the years, some of the leadership of the Green Party, particularly Caroline Lucas and Derek Wall, have strongly campaigned on issues that impact on the working-class. These and other key individuals, especially those in Green Left and the Green Party Trade Union Group, have supported working-class-led campaigns wherever possible including, for example, the recent McDonald's workers' pay strike.

Finally

This chapter illustrates the kinds of barriers that working-class people face in relation to engaging with environmental movements and organisations. As individual anecdotes, the stories do not provide fool-proof evidence of environmental classism. The incidents described may be interpreted in many ways and contrary examples can always be found. However, the lack of consideration or understanding of working-class lives described by myself and my research interviewees here is generally supported by the wider body of literature, as mentioned. Therefore, it seems environmental classism exists in these organisations and there is, consequently, a need to increase understanding of how to develop environmental organisations which benefit, include and respect working-class people. Before going on to discuss how to do this in the final chapters, the next chapter looks at the environmentalism that working-class people, themselves, initiate and lead.

6

Working-Class Environmentalism

When mainstream environmentalists think of winning over the working-class to their way of thinking, as described in the previous chapter, they are not seeing the working-class as already having an environmental consciousness, one which has been expressed for centuries. Middle-class environmentalists and their organisations have critically failed to recognise that working-class people have been waging environmental struggles long before their own environmental organisations and programmes came into existence. Working-class activists have, in particular, brought greater awareness of the environmental contaminants present in their own workplaces and communities. They have commissioned research and put pressure on companies and governments to adopt and enforce strict environmental policies and regulations, using the means at their disposal—educating, organising, collective bargaining, demonstrations and strikes. This chapter describes these activities, outlining and analysing the key contributions of working-class people with their own distinctive forms of environmentalism—the 'environmentalism of the poor', the environmental justice movement, and trade union health and safety campaigns. Firstly, though, I want to discuss the idea that environmentalism is a concern of only middle-class people.

© The Author(s) 2020   **139**
K. Bell, *Working-Class Environmentalism*,
https://doi.org/10.1007/978-3-030-29519-6_6

Environmentalism as a Middle-Class Issue

There is a prevailing view that poorer people are too preoccupied with meeting their basic needs for food, warmth and security to be able to think about, or be active on, environmental issues. This view has largely been informed by three theoretical perspectives that argue that environmental concern increases with affluence: the theory of 'post-materialism' (Inglehart 1977, 1990, 1995, 1997, 2000); the 'affluence hypothesis' (Diekmann and Franzen 1999; Franzen 2003; Franzen and Meyer 2010) and the 'environmental Kuznet's curve' (EKC) hypothesis (Yandle et al. 2002). Inglehart's post-materialist values theory asserts that, as societies grow more prosperous over successive generations, there is a shift in citizen values from predominantly materialist, focused on survival and security, to predominantly post-materialist, concerned with identity, rights and quality of life. The 'affluence hypothesis' similarly assumes a direct link between affluence and environmental concern, arguing that environmental quality is a public good for which demand rises with income. This theory claims that wealthier people 'not only have a higher demand for a clean environment, but [are] … more willing and able to reduce their standard of living in order to devote more resources to global environmental protection' (Franzen 2003, p. 299). The idea that affluence is a prerequisite for efforts to address environmental problems is also embedded in the 'environmental Kuznets curve' (EKC) hypothesis, that nations will see environmental improvements only after a certain level of national income is exceeded (Yandle et al. 2002). Partly as a result of these theories, until recently, it was widely considered that environmental awareness and concern occurred only among middle-class citizens in wealthy, highly industrialised and primarily Northern Hemisphere nations. The post-material values, affluence theory and EKC theories result in policies to grow the economy, which is problematic in itself, as has been discussed earlier, but these theories may also be basically incorrect for several reasons.

To begin with, the post-materialist and affluence theories clash with increasing evidence that poor people and poor nations are very active on environmental matters. In particular, low-income countries' representations at global events, such as the Rio Earth Conference in 1992, indicated that citizens of poor nations had higher levels of environmental concern than the post-materialist perspective recognised. Indigenous cultures have valued living in harmony with nature, with respect for its limits, for thousands of years (Pellow 2018). Furthermore, there is now rapidly accumulating evidence of environmental activism in low-income countries (e.g. see Martinez-Alier 2002, 2003; Pellow 2007). A number of quantitative studies have also challenged these post-materialist and affluence theories, finding that, globally, post-materialism and economic affluence are not consistently positively correlated with environmental concern (e.g. Dunlap and Mertig 1997; Dunlap and York 2008). Furthermore, the EKC hypothesis has been found to apply only in a limited set of circumstances (see, for example, Cavlovic et al. 2000).

My conversations with working-class people, as well as my own experience, also do not generally coincide with the post-materialist, affluence and EKC theories. Most of the working-class people I spoke to were very clear that they were concerned about the environment. For example, Dave said:

> *Yeah, clearly, I think anyone with a brain would consider that it's an issue … The key issue that I would see on a local level would be the level of pollution … when you consider what is a lovely city, a big spoiling aspect of it is quite frankly the traffic fumes … and the long-term effect that's having on people's health … It's as much a part of your quality of life, as a decent standard of living, you know, and a decent education and a decent health service. So I would say, it's something we have to struggle for.* (Dave)

It was not just the health aspects of the environment that concerned them, as several said how important nature was to them, as in the following excerpt:

> *It's good that people are using less plastic now … I think people see [on TV] animal's guts all screwed up with plastic bottles … and they think, what can I do? We have to think this planet belongs to everybody and we've all got a stake in it. I was really shocked when you see lakes filled with plastic and remote coastline with all this stuff on their beaches and what it was doing to sea-life, it really shocked me. That seemed like an absolute tragedy to me.* (Amy)

Others, however, did make some statements which fitted with the post-material, affluence and EKC theories. They argued that life is too difficult without having to think of the problems of the entire world. However, not thinking, talking or doing anything about a problem did not mean that they did not care about it. The conversations indicated that people feel it is not productive to think about the environment because there is nothing they can do, it is overwhelming, and it adds to their stress when they do think about it. For example, when I asked Ray, a taxi-driver, if he was concerned about environmental problems, he said:

I don't think about that kind of thing. I blot it all out. I have got a family and … kids to think about so I just focus on that … We should just make the most of life and not worry about things we can't do anything about. (Ray)

From the interview, it was clear that Ray saw the environment as something that was beyond his realm of influence. Later on in the conversation, Ray described his difficult economic situation where the recession had led to the loss of his former business, resulting in large debts and what he described as 'a nervous breakdown'. Overcoming these problems was all he was able to focus on.

Several of the other working-class people I spoke to also said that environmental issues were not a priority for themselves and other working-class people that they know, as exemplified in Sharon and Mel's words here:

A lot of people [on the estate] *are carers for other people, as well, so I think their lives are just so focused on the immediate rather than a bigger picture thing you know. It* [the environment] *isn't part of the conversations that I have with people. It is more kids and schools and housing.* (Sharon)

When I do think about it [the environment], *my main thoughts are that I should think about it more than I do … I would be lying to say that I thought about it because the day-to-day aspects of your life, just being able to pay your bills, find food, maintain your life, that tends to take dominance over the environmental aspects … people are aware that they're being screwed over in so many different ways but to actually do anything about that is another thing.* (Mel)

These examples might seem to fit with the post-materialist thesis and affluence hypotheses but, later in the interview, environmental concerns were expressed in each case. However, these concerns were sometimes different from the dominant environmental discourses around climate change and biodiversity. Knight and Messer (2012) found that environmental concern is complex and there are different aspects of 'concern'. Working-class people have often, but not always, had a different way of expressing their environmental concerns and a different focus. Whilst middle-class people in wealthy countries have focused on climate change and its remedy in the form of market mechanisms, technology and lifestyle changes, the working-class and the global poor have tended to focus on maintaining environments that are adequate for health and physical survival more immediately (see Bell 2014). For example, Julie said:

> It [climate change] *doesn't really bother me so much because I always think there has always got to be some sort of climate change. We have had ice ages and all these different ages that nature does anyway, so I think a lot of that is probably to do with nature doing its thing … I do worry about pollution … the companies that get away with flooding the waterways with their toxic wastes and things like that.* (Julie)

As mentioned in the previous chapter, working-class people are also less able to carry out those green activities which have a direct or indirect financial cost—buying organic food; purchasing long-lasting but expensive products; eco-tourism holidays and so on. Even enjoying nature can have a cost if we consider the social pressures involved in accessing nature, as Anita discusses here:

> *If we look at outdoor culture now, everyone's wearing clothes that are more expensive than we can buy. I just thought I couldn't go out into the countryside because I didn't have the money for the right kit and I would notice that I was just wearing a regular city jacket and everyone's got their mountain jackets or whatever and I would get wetter than everyone else and, if we're all getting wet together it would probably be fine, but I always felt like the sad one in the group with leaky shoes and just the wrong gear.* (Anita)

No matter how negatively or disinterestedly the working-class people I spoke to referred to environmentalism or environmentalists they all went on to talk about ways in which they are thinking about, concerned about, or taking action on, the environment (see later in this chapter).

The post-material values and affluence theories do not take into account different ways of expressing environmental concern. Some of the questions used in the studies on post-material values asked people if they engaged in green consumerism and environmental activism. These dimensions of environmentalism would not manifest highly among working-class people or poor people around the world, not because of a lack of concern about the environment, but because of all the barriers to their undertaking these particular activities, as outlined in Chap. 5. There is a general tendency to attribute environmental behaviour to dispositional factors (attitudes, values, etc.) rather than contexts (income, time, availability of greener choices). For example, studies of recycling have shown that wealthier White people tend to report higher levels of willingness to recycle, as well as actual recycling behaviour, than less well-off BAME people (RRF 2002, 2004). They then advocate information campaigns targeted specifically at the lower income BAME households. However, this fails to take into account the diverse living situations of the two groups, with Black people more likely to live in high-density accommodation with poorer recycling facilities.

Overall it seems that working-class people are just as interested in protecting the environment as middle-class people but there may be barriers to expressing it. Moreover, working-class environmentalism often goes unrecognised because it often has a distinctive form.

The Environmentalism of the Poor

The term 'environmentalism of the poor' was coined by Joan Martinez-Alier to describe the social struggles for the environment led by subaltern social groups (Martinez-Alier 2003). Martinez-Alier has shown how the subaltern classes, and the poor in general, often defend the environment in which they work and live because they are highly aware that it supports their livelihoods, wellbeing and survival. There are also many other reasons why it makes more sense to think of the working-class as environmentalists. Firstly, as described in Chap. 3, low-income and working-class social practices are generally much less environmentally damaging than the practices of the middle and upper classes. Poor and working-class people are often default environmentalists, automatically being kinder to the environment without conscious choice or activity because their, generally lower and more insecure, incomes tend to mean they consume much less. Since almost all consumption involves using resources and creating waste, they are environmentalists in deed, even if not in intention. As Joan Martinez-Alier argues, 'To assert that prosperity is conducive to the spread of postmaterialist values, implies forgetting the very material roots of prosperity' (Martinez-Alier 1995, p. 2). The roots of prosperity are the environment.

Another reason for thinking of the working-class as environmentalists is that, in poorer countries and among less well-off people, there is often a greater direct experience or perception of environmental degradation (see Chap. 3), which some analysts have suggested may be a key source of environmental concern (e.g. Dunlap and Mertig 1995; Knight and Messer 2012). This concern can be quite focused on the locality but, with increasing globalisation, working-class environmental concerns can also extend beyond the geographically immediate. For example, at the CLASS conference in 2016, an environmental think tank explained that they had been asked to speak to a group of Bangladeshi women in Tower Hamlets about climate change. The women were very concerned because their relatives back in Bangladesh were experiencing floods resulting from climate change, had lost their houses and were short of food. The women in Tower Hamlets were sending them money but also wanted to learn what they could do to help prevent further climate change (CLASS 2016). This illustrates how, rather than having more important things to worry about, poor people's and poor countries' problems are entwined with environmental problems. Working-class people are both metaphorically and literally at the coal-face of environmental deterioration because we tend to work in the most hazardous environments, live in the unhealthiest neighbourhoods, and are the least able to find individual solutions to all of this, such as changing jobs and homes. Therefore, working-class people have a strong vested interest in achieving an equitable transition to sustainability.

Furthermore, there are many examples of working-class environmentalism, both in relation to environmental preservation and in terms of campaigning on the environmental determinants of health and wellbeing. There is, in some respects, a hidden history of working-class environmentalism.

While the prevailing narrative of official and academic environmental histories has been that campaigns to protect wild spaces and establish national parks were initiated by middle-class activists, there are a number of examples, both in the UK and internationally, where this was not the case. For example, in the 1950s in Sydney, Australia, it was working-class activists who demanded conservation of native bushland and accessible green space. Goodall and Cadzow (2010), collating oral histories on the life stories of individual campaigners for this cause, trace how the local river and bushland environments were as important to working-class identities as their employment. The authors called for recognition of working-class commitment to saving, conserving and maximising access to green space.

Similarly, in the UK, working-class people in the 1930s had a major impact on our access to the countryside. In the eighteenth and nineteenth centuries, Enclosures Acts packaged up British common land and moved it into privately owned estates. Millions of acres of common land, which had been used by poor rural people to graze cattle and grow crops, were privatised, removing livelihoods, ways of life and often access to land for leisure purposes. Successive parliamentary demands for the right to roam on this land were made from 1884, all initially unsuccessful. Therefore, as part of a long campaign to gain access to the countryside, in April 1932, over 400 mostly working-class people participated in a mass trespass onto Kinder Scout, a moorland plateau in the Peak District. The event was organised by the Manchester branch of the British Workers Sports Federation, a subsidiary of the Young Communists League. The trespass was seen at the time, as a working-class struggle for the right to roam versus the rights of the wealthy to have exclusive use of moorlands for grouse shooting. Five men from Manchester, mostly working-class, including the leader, Benny Rothman, were subsequently sent to prison for breach of the peace and unlawful assembly for up to six months. The event had a far-reaching impact, widely credited with leading to the legislation passed by the Labour government in 1949 to establish the National Parks, the development of the Pennine Way and many other long distance footpaths, and securing walkers' rights in the Countryside and Rights of Way Act passed by Labour in 2000. Although the role of this trespass in the wider campaign should not be overstated (see Hey 2011, for a more complex and detailed account of the overall campaign), overall, the Kinder Scout trespass was a seminal moment in the struggle for public access to private land and access to nature. Most people in the UK do not know about the Kinder Scout trespass and the associated campaigns, taking the 'right to roam' for granted. Even so, the trespass has since been described as 'the most successful direct action in British history' (Roy Hattersley 2007). The Ramblers, now mostly a middle-class walking group, disassociated themselves from the trespass at the time, but now claim it as a victory.

The Environmental Justice Movement

As the above section indicates, poor, marginalised, working-class and other disadvantaged people have resisted environmental destruction for hundreds of years. However, the concept of 'environmental justice', sometimes instead called environmental equality, is relatively recent. The term has been used as a policy principle, as a campaign slogan and as a social movement agenda (Walker 2012). It is generally attributed to struggles in the US during the 1980s, when working-class BAME communities began to protest about hazardous and polluting industries being located predominantly in their neighbourhoods. This grassroots campaigning was largely occurring outside mainstream environmental activism, which was then mostly focused on the ecological concerns of middle-class Americans, such as the extinction of species and wilderness preservation. The 'spark' for the environmental justice movement is usually traced back to protests against plans to site a toxic landfill in Warren County, North Carolina. This landfill was intended to accept polychlorinated biphenyl (PCB)-contaminated soil that had been retrieved from illegal dumping of toxic waste along roadsides. The state of North Carolina considered a number of potential sites to host the landfill, but ultimately settled on Afton, a small African-American community in Warren County. The citizens spent years trying to convince the state officials of their concerns about the site, but frustrated and furious, they ultimately decided to lay down in the road in front of the 10,000 truckloads of contaminated PCB soil when it came to be delivered. The collective non-violent direct-action protest lasted for six weeks and there were over 500 arrests. It was described as the largest civil disobedience act since Selma, Alabama, in 1965, a key moment in the Black civil rights movement of the 1960s. One of the arrested activists that opposed the landfill, Reverend Benjamin Chavis, coined the term 'environmental racism', to articulate the perception of disproportionate siting of health-impacting facilities in predominantly Black communities.

This event was followed by a number of research studies which provided concrete data-based evidence of a significant correlation between the location of hazardous waste sites and the racial demographics of the county (e.g. Bullard 1983; US GAO 1983; UCC CRJ 1987). Robert Bullard's work, for example, showed that, in Houston, a city with a 25% Black population at that time, six of the eight city-owned waste incinerators and three of the four landfills were placed in predominantly Black neighbourhoods. Environmental justice activism and research then grew rapidly, including the development of a national network of activists which culminated in the 1991 First National People of Color Environmental Leadership Summit, held in Washington, DC. This meeting—organised and attended by more than 650 grassroots and national leaders, including the Warren County residents—heard a common complaint that their communities were targeted because of their racial and economic demographics. The Summit produced a consensus document which adopted *The Principles of Environmental Justice*—which included the right to freedom from ecological destruction; participation as equal partners at every level of decision-making; a safe and healthy work environment; and the cessation of the production of all toxins, hazardous wastes, and radioactive materials (First National People of Color 1991).

Following the environmental justice activities around the country, in 1992, the US government set up an Environmental Equity Working

Group and, later, the National Environmental Justice Advisory Council (NEJAC). In 1994, the Clinton administration passed Executive Order 12898 which focused on health and environmental conditions in Black, minority and low-income communities, instructing every national state department to develop an agency-wide environmental justice strategy and to ensure that any of their activities that substantially affect human health or the environment should be conducted in a way that does not discriminate on the basis of these demographics (Clinton 1994: Sec. 2-2).

At this point, the US was the most advanced country in the world with regard to specific legislation and policy on environmental justice. It has seen setbacks since then under the administrations of George W. Bush and Donald Trump, with its associated institutions losing financial support and legislative power. Also, attempts to introduce environmental justice legislation have been met by a continual backlash from businesses and the political right who have often perceived regulations as threatening their financial interests. Donald Trump's administration is proposing a 25% reduction in the Environmental Protection Agency's budget which will mean the loss of 3000 jobs and the entire Environmental Justice Office.

However, at its peak, the environmental justice movement was able to prevent the siting of toxic facilities in many working-class communities (see, for example, Pellow 2007). The notion of 'environmental justice' has also been taken up by social movements, and, in some cases, policy makers in many other parts of the world (see Bell 2014; Walker 2012). Walker (2012) notes examples of the term now being used in Europe, Africa, South America, Australia and South Asia.

In the UK, environmental justice, as a concept, emerged about 20 years ago, mainly via the academic community who focused on the geographical relationship between environmental burdens and demographics. As a result, after the US, the UK appears to have the greatest number of published research studies (in the English language) on environmental justice (Reed and George 2011). Environmental justice has also, at times, been debated by the UK government and policy makers (see Chap. 3). However, local working-class communities in the UK would rarely use the term 'environmental justice' (Bell 2008). Even so, there are struggles happening all over the country that could be defined as such, with working-class people frequently organising to defend themselves against toxic and inequitable burdens falling on their communities.

Many of my research respondents, all working-class, have been organising in local campaigns to address pollution, preserve green space, resist unwanted housing developments and more. However, their activities tend to not reach the media unless they are particularly creative or dramatic. For example, one group managed to get some publicity about the fly infestation caused by waste storage, described in Chap. 3. Paul explained how a few of the campaigners:

> … *chased* [the Mayor of City], *around … dressed as a fly* [laugh] *which was interesting* [laugh]. *It made the news … you've got to be creative you know, you've got to keep these people on their toes … If we notice that something is going wrong and it's not being addressed, we will go and camp outside their door and make an absolute furore until something is done.* (Paul)

Some have also been engaged in individual acts of environmentalism, waging their campaigns alone or trying to make a difference without any recognition or fanfare. For example, Jim and Ange said:

> *I felt City Road hasn't got a single tree in it until you get, you know, further down, so I went to a nursery and I picked up an Indian bean tree, and I … planted this tree, and it's now … the biggest tree in* [local area] *… I'm a big nature-lover. Why I live in the city, I don't know, but I always feel comfortable with nature, so it just seemed a nice thing to do for the community … Interestingly enough, what happened is that our neighbour, the husband died and it was very sad, I … put a rose tree there, and now that rose tree has somehow grown attached to the Indian bean tree, so, for the first time after seven years of planting this tree, the Indian bean tree is full of white roses.* (Jim)

> *I had the water tested* [for pollutants]. *So, basically, … I refused to pay the water bill, so, they took me to Court. The Judge ordered me to pay and they also ordered them to test the water and when they test the water they sent me the result and the results were what I expected, the water is pretty much polluted. They say they can't do nothing about it, so what I do, rather than not pay the water at all, I pay what I think it is worth, for the flush of the toilet and that's what they're going to have. No more than that … They tried to bother me again after that, but then I said 'Okay, sure, we go back to the Court with that test' and they left me alone … I'm not going to give them money for the water they give me that is polluted.* (Ange)

My interviewees' stories support other research and my own community development experience, over the years, which indicates that, in general, there are high levels of sustained community activity in disadvantaged areas. However, people in deprived communities need to see results in order to sustain motivation (Burningham and Thrush 2001; Brown 2002) and want to work on their own agendas rather than those imposed on them (CDP 1977; Dinham 2005). Sometimes their work on the local environment goes unnoticed. As one report concluded stating,

> It is important to recognise the hard work of those residents and activists who strive to change their estates. Often this is done quietly, without pay or recognition of financial costs to themselves ... The amount of time and effort these residents invest in their communities is often not recognised by agencies, who sometimes even take the credit themselves for it or disregard it. (Peace and Milne 2010)

As a community development worker, I was regularly engaged in supporting working-class people to address local socio-environmental challenges, for example, working with the Hartcliffe Health and Environment group. Struggles to protect and improve local environmental quality are happening in working-class communities across the country, sometimes noisily but often quietly and determinedly. These local struggles can remain hidden because the working-class people involved usually do not have the friends in high places that could help draw attention to their work—in the media, in government and in academia. This hidden struggle contributes to the illusion that environmentalism is the domain of middle-class people.

Trade Union Environmentalism

Some environmentalists consider that the trade union movement is largely irrelevant to the achievement of a sustainable society since, at least in the Global North, protecting wages has been the main focus of their activities. They can portray the working-class in the Global North as wealthy, overconsuming and selfish. There is also an overt or implied assumption that environmentalism will mean sacrifice for the working-class of the 'developed' world. Sarkar states, for example, that in the transition to sustainability 'first world workers have more than their chains to lose, they have their prosperity' (Sarkar 1999, p. 237).

The relationship between trade unionists and mainstream environmentalists has not always been easy, at best being one of distrust and suspicion and, at worst, of open hostility (Räthzel and Uzzell 2012). As discussed in Chap. 5 the conflict has often centred on the issue of 'jobs' with environmental movements accusing trade unions of defending jobs at any cost and trade unions perceiving environmentalists as putting nature before workers. For example, in the US some workers have used car bumper stickers saying, 'Are you an environmentalist or do you work for a living?' (White 1995). Trade unions have also been, at times, dismissive about environmentalism. One of the working-class people I spoke to for this research also made this point:

> With the TUC [Trade Union Congress], *if you say anything about climate change, you get accused of being part of the liberal metropolitan elite. Jobs are sacrosanct. It is not possible to talk about this with some people.* (Liam)

Trade unions have also sometimes supported environmentally damaging actions as evident in, for example, the trade union position on fossil fuel extraction from the Alberta tar sands in Canada. While environmentalists were warning that exploiting these tar sands will result in an increase of CO_2 in the atmosphere such that it will be 'game over' with regard to climate stabilisation, the prevailing view taken by the key trade unions was that extraction should go ahead since it supported jobs (Sweeney 2012).

Although trade unions may not always be on board with the environmental agenda, environmental movements are rarely subject to the same scrutiny when discussing the rift between unions and environmentalists. Environmentalists can do much to alienate workers, as the previous chapter describes, yet the blame for this rift seems to most often fall on the trade unionists with little self-reflection on the part of environmentalists. Rather than depict workers as pro- or anti-environmental, it is important to look at the circumstances that lead to their position. For example, a study of copper smelter workers in Montana, US (Mercier 2001), examined the various scenarios in which the local union pressed for or against stronger pollution laws. Mercier illustrates how workers in extractive industries, particularly in single-resource towns, are both physically susceptible to illness resulting from the pollution and economically vulnerable to industry cuts or closures and so have to balance their demands, sometimes emphasising environment and other times considering ongoing employment.

The ongoing social struggles that trade unions have to work on are immense, yet unions are focusing on environmental issues more than they are currently given credit for. On the topic of climate change, Hampton (2015) notes the myriad of related protest activities that trade unions have engaged in. Around the world, trade unions have been involved in climate change protests at national and international levels, for example, joining the mobilisations outside the international climate change talks at Copenhagen in 2009 and all those since, demanding government action on the environment and climate change. In all of these international climate mobilisations, the UK trade unions were present with their members, banners and slogans.

Trade unions are now taking on climate change as an area for the development of policy. Since the 1970s, they have been calling for a 'Just Transition' (ITUC 2015). The concept of Just Transition has now become part of global union discourse, though there is a great deal of variability in its interpretation, reflecting political preferences about the broader political economy (Stevis and Felli 2015). In 1997, the International Confederation of Free Trade Unions (ICFTU, now ITUC) included the term 'Just Transition' in its statement to the Kyoto Conference. At the 16th COP in Cancun in 2010, the concept of 'Just Transition' was used in the final agreement (see UNFCCC 2010).

Alliances are now being built between trade unions and environmental groups around the world (Stevis 2013; Sweeney 2012; Gingrich 2012). For example, in the US, the BlueGreen Alliance has formed out of an initial collaboration between the Sierra Club and the United Steelworkers union. In South Africa, environmental organisations are running training for trade unionists and collaborating in the 'One Million Climate Jobs Campaign' (Cock and Lambert 2012); in Brazil, the umbrella organisation for environmental organisations (FBOMS) has the national trade union, CUT, as a member of its management board (Räthzel and Uzzell 2013). The first international Trade Union Assembly on Labour and the Environment took place in Nairobi in 2006 under the auspices of UNEP (UNEP 2006) and, at each COP climate change international meeting, the number of unions represented has risen. During the earlier COPs, unions were largely observers, but more recently they have made strong demands for an international commitment to reducing greenhouse gas emissions (in 2006 at COP 12 in Nairobi) and produced an explicit position on international climate negotiations (in 2008 at COP 14 in Poznan). ITUC international conferences usually now include environmental leaders as speakers. There is collaborative work between the ITUC, the International Labour Organisation (ILO) and the UN Environmental Programme (UNEP) to promote 'green jobs' and a 'Just Transition' (UNEP 2008, 2011; ILO and UNEP 2012; Räthzel and Uzzell 2013; ILO 2018). The ILO (2018) have recently produced a report which states that action to limit global warming to 2°C will create enough new jobs to more than offset job losses of six million elsewhere. They predict that 24 million new jobs will be created globally by 2030 if the right policies are adopted and implemented.

Though trade unions and their members differ greatly in their views about climate change and the environment and the correct way to address these issues (Räthzel and Uzzell 2011), the notions of 'Just Transition' and 'green jobs' have been rallying slogans that have mostly been enthusiastically endorsed by all. To some extent this is because both have been left fairly open to interpretation (Stevis and Felli 2015). Some union leaders have, however, been wary: 'They can see a transition is underway since jobs are being lost, but they don't see what is being done to make it just' (Labor Network for Sustainability 2018). Although green jobs and Just Transition may be vague and problematic concepts in some ways (see also Chap. 4), at least they have acted as a point of engagement between workers and environmentalists to allow the opportunity for debate about how to have decent work without overstepping planetary limits.

The British Trade Union Congress (TUC) adopted a 'Just Transition' agenda in 2007 (TUC 2008). British trade unions have also taken up the related questions of fuel poverty and energy prices, and some have called for interventions aiming at the public ownership of industries, natural resources and publicly owned transport on environmental grounds. Successive TUC resolutions have linked public ownership with reducing carbon emissions and trade unionists also demanded one million climate jobs guaranteed by the state. The TUC has also run 'green camps' at Tolpuddle, the annual trade union commemoration event. At one point, the TUC explored the idea of a 'third sector alliance' on climate action, made up of trade unions, voluntary sector organisations and local community groups, among others (Hampton, 2015). There have also been a number of local union actions on climate change. In some organisations, workers have managed to get the right to a say over environmental decisions, time during working hours to progress climate issues, inclusion in audits and inspections of the workplace and the chance to make recommendations for change (Hampton, 2015). For example, in a project called JUPITER (Join Us People in Tackling Energy Reduction) at a brewery in South Wales, a UNITE union-led energy-saving project cut carbon emissions by 40% in two years by taking suggestions from the shop floor (South Wales Argus 2010). So, although my interviewee, Liam, felt that he could not talk about climate change in the TUC, as mentioned above, and some workers are wary, there is a great deal of activity to try to make these links within the union infrastructure organisations.

We should also not overlook the massive trade union contribution that unions have made to environmentalism in terms of health and safety. Climate change is not the only lens through which trade union environmentalism should be assessed. Though some might argue that working-class environmentalism is different from working-class occupational health and safety activism, I would argue that there is no difference, because of the centrality of health to environmentalism. Some of the most important environmentalism has been in the context of workplace health struggles. Activists have worked and lived with contaminating industries and were the first to see the toxic impacts. They were aware that the pollution they were exposed to at work would eventually seep into the wider environment. In a form of 'barefoot epidemiology', workers have, for centuries, exposed the toxic impacts of pollutants.

Environmental standards for toxic substances are seldom set prior to their industrial production, often developed only when illnesses among workers and exposed populations become too glaringly obvious to ignore. Although most people assume that the substances in use are safe, this is often not the case (see Chap. 3 and Lancet Commission, 2017). It has frequently been workers who have challenged this assumption, exposing toxicity, not only in the production process but in the products themselves. Because research was often employed to delegitimise workers' concerns about environmental contaminants, the workers have had to carry out their own studies.

Many trade unionists have lost their lives in the battle to expose environmental harm. For example, Chico Mendes of the Rubber Tappers Union of Brazil was eventually killed to put a stop to his campaign (Revkin 2004). Even if not directly killed, drawing attention to the ecological damage or human toxicity of production or admitting illness caused by it often meant the loss of a precarious job and livelihood.

Even so, there has been a long history of trade union negotiation, political and pressure group activity, as well as outright protest and resistance, centred on health issues in the workplace, particularly evident in records of local and unofficial strike action in the so-called dangerous trades. In 1902, Sidney Webb and Beatrice Webb commented that

> *In the trade union world of today, there is no subject on which workmen of all shades of opinion, and all varieties of occupation, are so unanimous, and so ready to take combined action, as the prevention of accidents and the provision of healthy workplaces.* (Webb and Webb 1902, p. 35)

The 'union safety effect' is very real and pronounced. One study found UK workplaces with an active union safety system had an injury rate half that of workplaces without these systems (*Hazards* 99). Another UK study found that the proportion of employees who are trade union members has a positive and significant association on both injury and illness rates. A UK government investigation found the savings to society stemming from the activities of union safety reps—fewer sick and injured workers—run to hundreds of millions of pounds every year (*Hazards* 97). Dave, one of the active trade unionists among the working-class people I spoke to in my research, gave a very personal example of how the UK Health and Safety at Work Act 1974, gained through workers' struggle, had impacted him:

> *I've been, on many occasions, a shop steward in my places of work. I think the greatest achievement for working people, was the implementation of the Health and Safety at Work Act. Now it's not perfect, and it's only really enforceable in a reality where there is a strong union organisation to enforce it. But, in terms of my working experience, where I began in heavy industry, labouring jobs, engineering, at a time before that Act became effective, the accident rate was actually phenomenal. I mean, personally, before the Act, I nearly had three fatal accidents, and the management's attitude to that was complete and utter indifference.* (Dave)

Trade union struggles, through their focus on health and safety in the workplace, have necessarily also been environmentalist. There are numerous studies on byssinosis, silicosis, asbestosis, 'miner's lung' and other conditions that describe trade union activities around environmental issues in the workplace (e.g. McIvor 2016; Long 2011; Mills 2010; McIvor and Johnston 2007; Bufton and Melling 2005; Bowden and Tweedale 2003; Tweedale 2000). The trade unions were central to campaigns against unhealthy work conditions. For example, in the UK, the Women's Trade Union League (WTUL) (1899) led a campaign against the use of lead in the manufacture of pottery. The lead used in the glaze was causing blindness, convulsions and death among the mostly female workforce. There were also impacts on unborn children. The Factory Act Annual Report of 1897 stated that, out of the 77 women suffering from lead poisoning during that year, 15 were childless; 8 had 21 still-born children; 35 had 90 miscarriages and 40 of their children had died in infancy as a result of convulsions. Ada Nield Chew, a working-class woman dubbed 'England's forgotten suffragist', was a key leader in the WTUL campaign to ban lead in pottery (Chew 1982). This is just one example of how concern for environmental health at work had impacts on wider society, with the population eventually no longer eating and drinking from lead glazed plates and cups as a result of the public and political attention that the campaign brought.

Some of these campaigns—for example, the Trades Union Congress (TUC) and Scottish Trades Union Congress campaigns on tuberculosis—resulted in the illnesses found among the workers eventually being recognised as an occupational disease (McIvor 2012). Union pressure also led to the establishment of Joint Health and Safety Committees across many UK industries in the 1960s, resulting in significant improvements in health and safety standards (McIvor 2016).

The struggle for recognition of occupational disease is well illustrated by the radium dial painters' story outlined in Chap. 1. This and the struggle to achieve recognition of lead poisoning, coal workers' pneumoconiosis, asbestos-related diseases, silicosis and many other diseases raise questions about the roles of academics, scientists and industry. Abrams (2001) explains the delay between knowledge of an environmental condition causing illness and death and anything being done about it in terms of the historically close relationship between science and industry. Scientists often follow the demands of the establishment. One of the most vivid illustrations of this is an extract of a letter written by a top asbestos industry executive to a prominent medical researcher which stated: 'It is our understanding that the results obtained will be considered the property of those who are advancing the required funds, who will determine whether, to what extent, and in what manner they shall be made public' (in Abrams 2001). The researcher agreed to this stipulation and, thereby, contributed to the long delay in recognising asbestos as a major occupational and public health hazard.

Official strikes have not been common in health and safety campaigns, but there were some. For example, in 1949, more than 5000 miners in a town called Asbestos in Quebec, Canada, went on strike against excessively dusty working conditions which they considered was undermining their health (McIvor 2016). Short unofficial strikes were much more common, though less visible in the records (McIvor 2016). There were numerous 'wildcat' strikes and spontaneous 'walkouts' on health and safety grounds at the local level. At times, deindustrialisation, alongside factory and mine closures, diverted trade union attention to other issues, such as maintaining a job and adequate wages.

However, McIvor (2016) describes how there has, at times, been some discomfort with occupational health and safety because of entrenched macho work cultures, where maximising earnings and taking risks are exalted and protective forms of behaviour pilloried as 'effeminate' (McIvor 2016, p. 74). Because men are socialised into high levels of danger and risk, they do not want to be seen as 'weak'. Pragmatism, fatalism and conformity prevail in these high-risk cultures.

However, while there may be some macho bravado involved, there are also strong economic incentives to keep quiet about health and safety. For some time in the UK there has been insecure work and a decline in industry, often in areas where there are few alternative employment opportunities. The risks of contracting workplace environmental illnesses can seem very far distant in the future while the need to maintain a job and adequate wage is a problem that must be addressed now. McIvor (2016) noted that those working in dangerous jobs often considered high wages as compensation and there were even actual 'danger money' payments until the 1970s. Despite this tension, in the later 1970s, the TUC began mass training of health and safety representatives—around 80,000 attended trade union health and safety courses between 1974 and 1982 (McIvor 2016).

In North America, a number of authors have also described how their environmental and labour history has been interwoven (in MacPhee 2014), Drawing on contemporary newspapers, union minutes and reports and the strike and lockouts files of the Department of Labour, MacPhee (2014, p. 123) describes a 'distinctly worker generated environmentalism born largely out of a class analysis of health and disease under capitalism'.

However, Barca (2012) argues that, despite their successes, unions have failed to be effective on environmentalism in three main aspects—firstly, by focusing on the site of industrial production, leaving out the question of the wider pollution of society, locally and globally; secondly, through insufficiently highlighting the connections between wider environmental and health and safety grievances; and, thirdly, through faith in a productivist paradigm based on economic growth as a means of addressing environmental and social problems. Taking into account the societal factors that support these failings, these are important critiques and the issues raised may be an agenda for trade union progress and will be further discussed in the final chapters of this book. These failings may not be quite as pervasive as Barca seems to portray, though, since there is some evidence that trade unions have been effective on the first two aspects, in terms of making connections with the wider environment.

MacPhee (2014), for example, perceived that trade union concerns about industrial pollution in the workplace sometimes led into a wider environmental consciousness outside the bounds of the workplace. Workers often developed links with local communities, involving them in their epidemiological studies or strikes over environmental health issues, and forming campaigning coalitions (MacPhee 2014). As an illustration, MacPhee describes how, in the 1960s, workers and residents in Yellowknife, Canada, suspected that the arsenic released as a by-product of local gold mining was responsible for the increase in local cancer rates. Their suspicions were dismissed by government officials until 1966, when the National Health and Welfare office discovered that arsenic levels in Yellowknife were ten times higher than the permitted threshold and that inhabitants were suffering from cancer as well as respiratory and nervous disorders at elevated rates. This report was held back from the public and the workers until 1975. The workers, allied with federal and regional First Nations organisations, then conducted their own joint study of their members to assess the health implications of arsenic exposure. Some of the samples they collected contained arsenic levels 50 times above the World Health Organization's designated 'safe' standard. One of the University of Toronto biologists involved declared Yellowknife to be probably the most severely arsenic-contaminated area in the world. As a result of these revelations finally, in 1978, the Canadian government announced its intention to create stricter regulations surrounding arsenic emissions from gold mining at a national level (MacPhee 2014).

It is important to recognise these trade union- and worker-led health and safety campaigns as environmentalism. As MacPhee points out, separating environmental activism and occupational health and safety struggles is surrendering to middle-class conceptions of environmentalism (2014). He proposes an expansive definition of working-class environmentalism which encompasses struggles over pollution at the point of production.

Finally

Working-class people have been, and continue to be, environmentalists. They are often impacted first, most directly and most severely by environmental degradation. The working-class are, therefore, integral to environmental activism. Cost-cutting industries have often obscured the health impacts of production, generally promoting their activities as safe. The proof that they are not safe often begins with dead and dying workers and local communities. Throughout the history of workplace and community health and safety we see fatal but preventable accidents, company denial of health hazards and the covering up of problems to preserve the public reputation of the company. Though trade unions and communities have sometimes prioritised jobs over health, they should not have to make such a choice, as discussed in the following chapters.

It is important that working-class people and their organisations are recognised as environmentalists if we want to see a transition to sustainability. Working-class environmentalism may seem to be overly focused on the human and the local. It may seem too anthropocentric to focus on environmental justice and human health and safety, rather than ecological justice, as this seems to imply a concern about nature only in as far as it affects human beings. However, as mentioned in Chap. 1, a healthy human environment can also lead to a healthier environment for other species. If we consider that humanity is dependent on the environment and the environment is made up of complex ecosystems that need all their constituents in order to remain stable, then we cannot afford to harm those systems or their inhabitants. Far from being no more than a source of raw materials and repository of waste, nature can have an aesthetic, moral, health, spiritual or scientific value for humans. There is no need to separate, and thereby create an unnecessary conflict between, middle-class concerns about wilderness and climate change and working-class concerns about human health, safety and economic survival. As the trade union slogan goes, 'a victory for one is a victory for all!'

Finally

7

Explaining Environmental Classism

The problem of environmental classism described in this book is socially produced and is, therefore, preventable. But we can effectively prevent it or address it only if we understand the causes. There are a number of potential causal explanations, operating at different levels. Both structure, the persistent patterns of social relationships, and agency, the self-directed actions of individuals, create and reproduce class and classism. The evidence presented in this book indicates seven broad explanations for environmental classism which have both an individual aspect and a structural aspect. These include class power imbalances; classist discrimination; income inequality; mutual hostility between workers and environmentalists; the separation of environmental and social aspects of life; the focus and composition of the mainstream environmental movement; and limited democracy. These explanations are interconnected and entwined but discussed here separately. I will also argue that, if we dig deeper, it seems there is one underpinning explanation: the capitalist economic system, that is, private ownership of most of the means of production, with economic outcomes resulting from market mechanisms, such as price signals, rather than through government planning. Capitalism's defining features of profit, competition and a class-based system of

© The Author(s) 2020   **163**
K. Bell, *Working-Class Environmentalism*,
https://doi.org/10.1007/978-3-030-29519-6_7

domination and subordination are, I believe, the central problems that need to be addressed.

Class Power Imbalances

The economic, political and social powerlessness of the working-class seems to be an obvious explanation for environmental classism. Powerless communities are less able to prevent or remove environmentally harmful facilities and processes or to access environmental resources. Industry and government may choose to locate hazards in working-class communities because they are the least able to mount an effective opposition due to a lack of finances, time, skills, contacts, representation and information. Middle-class people are more likely to be able to employ lawyers, lobbyists and experts to defend their neighbourhoods against the siting of toxic facilities and to support their cause when they fight for environmental and service improvements. 'Middle-class capture' (see Chap. 4) prevails. Gal (1998) suggests six channels of influence that permit 'middle-class capture'.

The first is via the formal political process or the 'electoral channel'. With their higher than average propensity to vote, the politicians listen carefully to the demands of the middle-class and are, therefore, more likely to devise policies and services that appeal to middle-class voters and avoid those that would be unpopular with the middle-class. Studies show that middle-class people are more active in trying to influence their political representatives (e.g. James 2007) and more likely to complain about public services than working-class people (e.g. Carvalho and Fidelis 2009).

The second channel of influence identified by Gal is the 'organisational channel' (1998, p. 47), with middle-class organisations exerting power in the development of policy, particularly at the local level, including land use planning and neighbourhood conditions (Hastings et al. 2014). A lack of time and energy can be barriers to working-class organisation. In recent years there has been even less control over free time, as people are engaged in part-time working on casual contracts where they are sometimes told only at the beginning of the day whether they will be required to work that day or not. Sometimes the strengths of middle-class campaigns are anticipated in environmental decision-making, as discussed in Chap. 4.

Gal's third channel, the 'knowledge channel', highlights middle-class influence through knowledge acquired through their education and occupations and their 'cultural capital', that is, 'their ability to communicate this knowledge in ways deemed appropriate in the context of service provision' (Hastings et al. 2014, p. 208). The fourth channel relates to the 'dominant role' of the middle-classes in the media, influencing how issues are framed (Gal 1998, p. 47). Gal's fifth channel of influence is the threat of 'exit'—the possibility for middle-class groups to choose the private market over public provision which, if occurring en masse, could threaten the viability of the service itself. Finally, Gal asserts there is the 'bureaucratic channel' which highlights how public agencies are run by the middle-classes and therefore their decisions will be based on their class affiliation and greater empathy with other middle-class people so that they 'serve as the guardians of the interests of the middle-classes' (1998, pp. 48–49).

Environmental disparities and inequities can occur because companies and governments take advantage of the relative disempowerment of working-class people in relation to these channels of influence. It can also occur because the companies and governments are merely following the path of least resistance, or because the inclusion of working-class people is simply not on their radar.

When working-class people are under-represented in political and administrative bodies they are less able to influence decisions on environmental planning and services. For example, the current membership of the Committee on Fuel Poverty is dominated by energy supply companies, rather than by the fuel poor who would be very motivated to eradicate this problem. On their current webpage (Committee on Fuel Poverty 2017), the Chair is described as having had 'a number of leading managerial and operational roles in multi-national oil companies in the UK and around the world'. Another member is ex-CEO of Npower and another is Chief Executive of Energy UK, the trade association for the UK energy industry representing over 90 suppliers and generators of electricity and gas. While there are a few directors of relevant charities, such as the Chief Executive of National Energy Action, a national charity seeking to end fuel poverty, the current board is predominantly made up of people who supply energy and sit on other boards.

Working-class people need to have positions in such organisations. They also need to be supported in their position so they can have influence. When working-class people are involved in such bodies, it can be tokenistic and frustrating, as discussed in Chap. 4. They are disempowered by processes which sometimes only allow reaction to decisions that have already taken place or minor issues. For example, planning rules tend to only allow opposition on self-interested grounds or technical issues, rather than the discussion of wider issues, principles and emotions. Sometimes, seemingly inclusive participation processes may be disguised attempts to control dissent and to legitimate decisions that have already been made. When working-class people try to alert the authorities to potential dangers, they may not be listened to when what they are saying is not convenient. Ange highlighted this stating:

> *Rather than look after us, they will look after big companies which they have absolutely no* [sigh], *no consideration at all for who's living here and the potential harms they can do with their business … When I told the* [Council Elected Members] *about, you know, can you investigate this incinerator, if it's safe. They didn't acknowledge much, and they didn't come back to me with a proper reply … 'Money talks'. So, if you have the money you have more value and people listen to you more. Hold on a minute! It doesn't work like that, you know, I'm a human being. Is this right? In my opinion it's not right, 'cause I'm not less valuable than someone who's got the money.* (Ange)

The kind of indignation on class grounds expressed by Ange is quite unusual. The concerns that residents express in an area do not necessarily link to class issues, according to my experience as a community development worker (see also Shaw and Mayo 2016). It is often the safer agenda that people propose, focusing on what can be seen in the here and now, rather than raising the issues that need to be addressed to really make their lives better, such as a redistribution of wealth and the elimination of toxic production. In addition, concerns raised often reflect dominant norms and debates in society at that time (Taylor and Wilson 2016) and this is why the middle-class capture of the media is important.

Over a lifetime of not being listened to or taken seriously, working-class people can give up. They can go along with the way things are, ignoring any internal or external messages that there is a problem that requires a response. Decades of 'learned helplessness' experiments from the field of psychology show that, generally speaking, when humans or animals endure repeatedly painful or otherwise aversive stimuli which they are unable to escape or avoid, they learn to stop trying to avoid the situation, even when the opportunity to leave presents itself (e.g. Seligman 1972). McGarvey (2017, p. 48) describes how in low-income communities there is 'a pervasive belief that things will never change' and that this arises in part because the enthusiasm to be active in communities quickly dissipates when people continually come up against walls when they try to change things.

This means people can designate certain environmental and economic problems as beyond their control, even when they may not be. Many of the working-class people I spoke to had resigned themselves to their local environmental problems or come to see them as beyond their control. One of the women I spoke to said that she was aware of the dangers of a high-tension pylon at the end of her street but,

I don't think about it. I have to say I don't. I see them because they're at the end of my street. I've got one right opposite facing me if I'm looking to the right, but I don't ... I don't look at it and think 'danger!'. I don't look at it and think it could be a problem, even though I know it could be. I don't look at it and think that. Maybe because it's about, like, how do you change it? How do you get rid of it? Do you know what I mean? Cos if I can't ... it's like the saying, isn't it, change the things you can and try and forget about the rest of it [laugh]. (Sharon)

Though not necessarily due to this environmental issue, Sharon became seriously ill with a life-threatening illness shortly after this interview. To some extent, her denial of the potential health risks in her local environment may be the best way of maintaining her mental health because she may not have the money available to move or the time to launch a successful campaign. Acknowledgement of the risk would merely cause her stress. According to my experience as a community development worker, a middle-class person would probably try to change a harmful situation that could potentially impact their health, however difficult. Feeling powerful in the world enables individuals to choose from a much wider range of possible solutions to their problems.

Therefore, working-class communities tend to lack power in relation to governments, businesses, and middle-class individuals and their organisations. This imbalance of power might be partially offset by the support of environmental campaign groups. However, as we have seen in Chap. 5, the mainstream environmental NGOs often do not focus on the issues that are important to working-class communities or engage with them in ways that are empowering and inclusive.

Power differentials explain environmental inequalities and inequities on the basis of class, but it is also important to consider what lies beneath these power differentials. The 'middle-class' capture notion helps us to understand that middle-class people are advantaged environmentally because they vote, get organised, have the kind of knowledge that is valued by elites, can communicate their knowledge in a way those elites value, and because they control the media and bureaucracy. However, this explanation, alone, would imply that the working-class just need to be helped to organise to gain improved environments and environmental services. This does not take into account the economic pressures that working-class communities face which can make it harder to get involved and harder to say 'no' to coveted jobs; Section 106 funding (private community investments, paid by local developers); and cheaper houses in environmentally degraded areas. Working-class communities can be more vulnerable to accepting toxic or otherwise unwanted facilities because they are more in need of the potential economic benefits that they claim to bring. Wealthier areas are much less likely to need any individual enticements or public facilities that may be offered. Working-class communities are particularly in need of the potential jobs, forcing them to choose between unemployment and the health and safety of themselves, their families and their communities. In reality, the claims about the hazardous facility bringing new jobs for the local community are often exaggerated or even false (Bullard and Johnson 2000), and compensation schemes which might appear to benefit the community can even increase the siting of toxic facilities in these areas. When there is a desperate need for income, offers of compensation can weaken or divide resistance. Lacking ownership or control over the means of production, the working-class are at a structural disadvantage. This situation can be beneficial for governments which do not have to take their views into account, so the disempowerment of particular groups is integral to the functioning of a capitalist system.

Classist Discrimination

'Discrimination' is the explanation for environmental classism that focuses on individual and institutional attitudes, actions and policies. In the US, Black and other minority ethnic academics and activists often cite 'discrimination' as the reason for the location of toxic facilities in their communities (e.g. Bullard 1983, 1990). According to this view, discrimination can be inferred from studies which show the socio-spatial correlation between particular demographic groups and the location of polluting facilities or otherwise unhealthy environments, without the necessity to prove discriminatory intent. Some analysts argue that this is insufficient evidence because these studies only provide a snapshot analysis and do not look at the social make-up of the neighbourhood when the facility was first established (e.g. Been 1994). It could be that people from low-income and other disadvantaged groups have moved to a location near the facility because the housing is cheaper. It could also be that the facility has been placed there because the land is cheaper. This kind of analysis argues that correlations of low-income communities with toxic facilities are the result of rational government and corporate actors simply weighing up costs and benefits, finding cheaper land in poorer areas, rather than intentionally discriminating.

Many other forms of discrimination in the UK are prevented or reduced by equalities legislation but class is not considered a 'protected characteristic'; that is, it is not covered by that legislation as discussed in Chap. 2. Even so, discriminatory intent would, in most circumstances, be perceived as immoral and so overt statements or acts are usually suppressed or hidden. But this does not mean class-based discrimination is not happening as Chap. 2 and other sections of this book illustrate. The polluting industries have repeatedly stressed that their siting decisions are not based on the demographic make-up of the proposed location and they can almost always offer some alternative justification for their decisions (Kiniyalocts 2000) but this does not mean they are not discriminating.

Discrimination does not have to be 'deliberate', in the sense of being fully conscious. We know that discrimination can occur without deliberate intent, such as through a lack of understanding of equalities issues, often embedded in historic, cultural and institutionalised practices. For example, institutional discrimination includes when environmental policies distribute the costs of controls in a regressive pattern, imposing disproportionate burdens on those with lower incomes (as discussed in Chap. 4). Furthermore, the institutional classism inherent in wider society will impact on environmental inequities, for example, where professional employers discriminate on the basis of class, restricting access to well-paid jobs (as discussed in Chap. 2).

Discrimination seems to work in conjunction with economic forces so that focusing on individual acts of intentional discrimination or even collective forms of institutional discrimination usually ignores the structural factors which create wealth and status inequalities. Whilst taking into account that low income is not synonymous with being working-class, as discussed in Chap. 2, working-class people are generally less likely to be able to earn the higher incomes necessary to afford to live in healthier environments. Low income reduces choices about where to live or which jobs to take, so that it can become impossible to avoid toxic facilities, whether located on the basis of discriminatory intent or not. People with low incomes find their mobility limited through financial constraints which link to educational and employment disadvantage, as well as urban segregation and housing discrimination. The larger societal contexts limit individual choices. Therefore, it is important to look at social conditions, rather than just individual decisions, and to understand how these constrain individual choices.

Dover (2016), in his work on 'microaggressions', highlights the importance of not trivialising class exploitation by reducing it to class chauvinism. Individual classist opinions and attitudes are just one aspect of oppression but, fundamentally, the problem of classism is that it enables class exploitation. Looking at the underlying drive for discrimination, it is apparent that capitalism promotes hegemonic classist discourses and values that enable discrimination and exploitation. We are socialised and conditioned to believe in meritocracy because, otherwise, it would seem unfair and upsetting that some people are allowed to have more of what is desirable simply as a result of privilege. Although classism is similar, in many ways, to other kinds of oppression, such as racism, disablism and sexism, arguably, it is more fundamentally based on the economic system itself. Hence, within capitalism, the working-class have a purpose—to produce as cheaply as possible. They are a factor of production and, at times, an expendable one. This exploitation requires some degree of dehumanisation which facilitates mistreatment.

The dominant classes—the owning and middle-class—tend to control the distribution of knowledge, thereby determining the 'truths' and the norms that govern society. This socialisation process leads people to believe the negative myths and stereotypes about the working-class and the positive myths and stereotypes about the middle-class, even among members of those groups. Therefore, even working-class people, themselves, can sometimes believe they lack intelligence and are inarticulate, incompetent and lazy, as discussed in Chap. 2. They can also tend to view those in the dominant classes as intelligent, ambitious, confident and making good leaders. During a lifetime of being told in countless subtle and less subtle ways that these stereotypes are true, people often become resigned to their situations. They can come to believe that they deserve what they have and nothing more; that this situation is justified and natural; and that their low incomes and low status are the result of their own deficiencies. However, there is nothing natural about discrimination or the inequality that both drives and is driven by it as the next section discusses.

Income Inequality

Environmental classism could not happen without the wider inequalities in society that produce social class. Chapter 2 explained the extent of these inequalities. Clearly, if some have more resources than others, they can access more opportunities and move away from environmental harms, if necessary.

However, inequality contributes not only to environmental classism but also to environmental harm more generally. A number of studies have investigated the link between inequality and various aspects of sustainability, finding correlations between inequality and worse environmental outcomes in terms of environmentally friendly behaviour (e.g. Dorling 2010a, 2010b); biodiversity (e.g. Pandit and Laband 2009); energy consumption (e.g. Baek and Gweisah 2013); industrial pollutant waste (e.g. Jun et al. 2011); air pollution (e.g. Torras and Boyce 1998); the number of green technology patents (e.g. Vona and Patriarca 2011) and levels of consumption (Wilkinson and Pickett 2009).

In particular, it is now being recognised that the most unequal societies consume more. For example, Dorling (2010a, 2017) has described how higher levels of inequality in a country tend to correlate with a greater consumption of meat, water and flights; as well as the production of more waste and a larger ecological footprint. His statistical analysis indicates that in more equitable affluent countries, such as South Korea, Japan, France, Italy and Germany, on average people consume less, pollute less and emit less carbon per capita than people living in the more unequal affluent countries such as the US, Canada and Britain. A 2016 study and report by Oxfam also found this pattern, with greater consumption occurring in more unequal affluent countries as a result of wasting energy, excessive heating of homes, driving larger cars and driving them more, flying more, choosing larger houses and shopping more.

Overconsumption can be explained as 'emulative consumption' (Veblen 1994 [1899]) where, in an unequal society, life becomes a battle for respect with material wealth being a means to gain or maintain that respect. The wealthy can afford to overconsume, but their habits influence the less well-off who then also often aspire to a similarly affluent lifestyle. As Ange and Sharon, two of the working-class women I spoke to, admitted,

We buy things that we don't really need. We just buy, we buy because the attraction when you go to the shop, they make it so attractive that you just say 'oh I

want to buy this.' They give you the strong desire that you need to buy and if you don't, you feel, you feel poor. (Ange)

It's such a throw away culture now and, yeah, it's not good … The pressure to have these things on the TV when they're doing all their adverts and … like with the gold and the latest trainers and having the latest fashions and that throw away culture, a lot of it is 'cos their self-esteem is low so they've got to be out there and be somebody or act the big, you know and 'cos they're looked down on so much. (Sharon)

Inequality generates a great deal of insecurity, advertising plays on that, and we tend to buy more than we really need simply to feel better about ourselves and to impress important others so that they think we are good enough. Therefore, we tend to buy status-symbol items such as new cars, fashionable clothes and whatever else will signal to our peer group that we are 'doing well' financially. The need for social approval and to avoid rejection is very strong in humans (Weir 2012). In some cases, these goods are required because others have them—when everyone is driving a car or using a mobile phone, it is expected that you will use them too and it will be unpleasant/difficult for you to try to travel or communicate in another way. In order to participate in society, we have to spend money on these, so-called, 'positional goods' (Hirsch 1977).

Capitalism has a tendency to polarise income and wealth inequalities and dominant discourses blame and shame those who seem to be less well-off. With its focus on profit and tendency to periodic crises, inter-mittent cost-cutting is necessary which creates unemployment, underemployment and low pay, as my previous work has discussed (e.g. Bell 2015). We have seen throughout this book how the class divisions that occur in a capitalist society favour some and marginalise others, with benefits and costs being distributed according to income and status. With the market deciding how commodities are allocated, there is no rational prioritisation according to need under a capitalist system. For example, it has recently been reported that approximately one in ten adults have bought or inherited at least one second home (The Resolution Foundation 2017). Meanwhile, 1 in 200 UK adults is homeless, rising to one in 25 in some areas of the country (Shelter 2017). Working-class people are sometimes not able to obtain even the most basic goods and services that they need because access is determined through sufficient money, credit or status. This occurs while the wealth created by working-class people is siphoned off to elites, enabling them to amass vast fortunes and to overconsume.

Mutual Hostility Between Workers and Environmentalists

As outlined in Chap. 6, trade unions and environmental movements have sometimes failed to support each other. In the Agenda 21 proposals from the 1992 Rio Earth Summit, trade unions highlighted how environmental issues are bound up with health and safety issues and how workers should be involved in decision-making around sustainable development, in the work environment and the production process. However, thereafter, some trade unions continued to focus on workplace health and safety, rather than on environmentalists' priority agendas, particularly climate change. This makes sense since few others are focused on the workplace environmental agenda. More recently, as Chap. 6 describes, trade unionists have taken on the classic environmentalist agenda of climate change. However, at the local level, workers have often found mainstream environmentalism to be a threat, not only to their jobs, but also to their identities (Barca and Leonardi 2016).

Environmentalists are sometimes seen to be part of an elite and there is now a great deal of hostility and mistrust of the elite. For example, some of the working-class women I spoke to said:

The whole country is going down-hill now. It's being run by a bunch of public-school kids. They are going off on their jollies, spending our money, and everything around here is falling apart. (Joan)

The ones in charge are only thinking about squirrelling away money on their tax havens, ripping off their expenses, having affairs and harassing women, and who they are going to bomb next. They don't care about us or anyone else. They are not even thinking about us. So I won't be told by the government or any of those posh people what I should be doing to meet some target they've got this week so they can get their promotion or whatever. (Sheila)

The companies cause a lot of it [the environmental problems]. *Probably the governments as well because so many of them are quite corrupt in a way that they will accept these backhanders to let the companies do what they want … The governments are letting them get away with it because they are getting backhanders.* (Julie)

In Chap. 4 we saw how environmental organisations tend to start with their own priorities and how they tend to lack knowledge about, and experience of, engaging with working-class communities. Even with the more working-class, environmental justice movement, as Barca and Leonardi (2016) point out, 'work and its complex relationship to environmental concerns is probably the least examined aspect of environmental justice struggles and of environmental conflicts' (Barca and Leonardi 2016, p. 62). The movement for green jobs should bring trade unionists and environmentalists together but, as discussed in Chap. 4, there is some scepticism in terms of who gets the jobs and whether they are just or exploitative.

The underlying rationale for this rift is supposedly that trade unions want jobs (which require growth) and environmentalists want to live within planetary limits (which require abandoning growth for its own sake). However, trade unions may need to question their assumption that growth equates with more jobs. Richard Douthwaite (1999) challenges the idea that growth benefits workers, arguing that as growth and profits have gone up for companies, wages have gone down for workers. The 'treadmill of production theory' points to capitalism's tendency to replace costly labour with technology. This issue of technology taking jobs could be a point of commonality between workers and environmentalists, as discussed in Chap. 8. Presenting the issue in terms of 'jobs versus environment' deliberately distracts attention from the fact that capitalism destroys both jobs and the environment. When solutions are framed with workers included in solving the problem of environmental degradation, then the confrontation between workers and environmentalists can be bridged. At the moment, this is not occurring to the extent that it could, but Chap. 8 explores how these two important social movements could better align with and support each other.

Separation of Environmental and Social

Some of the division between mainstream environmental organisations and organisations which mobilise for social justice, including trade unions, can be traced back to a more general separation of nature and society. This compartmentalisation of the human and environmental realm is evident in institutions and discourse round the world which deny or minimise the interrelationships between humans and the ecological system as a whole. The division is embedded in Western culture which has a long history of dualistic thought which promoted a severance from nature, with human beings placed above because they possess mind and soul, and nature situated below, existing only to serve human needs. We have a resulting tendency to think of ourselves as if we are not part of nature.

Although most of those I interviewed felt connected to nature, it was by no means unanimous. For example, Julie remarked:

> *I don't go out. I am a sort of home body. I am not a nature person … I am not a country type person … I am a towney … I have got this huge garden and it is enough that I have to mow it!* (Julie)

While deep ecologists and some indigenous groups believe that humans and nature are fundamentally one and the same and this perception of separation results in harming nature and ourselves (for a debate on this, see Vucetich 2017), in the Western world we are tending to become more distanced from non-human nature. For example, recent research indicates that British people are becoming increasingly detached from wildlife, the countryside and nature. Seven out of ten people in the survey felt they were losing touch with nature, a third said they did not know enough about the subject to teach their children about it and one in three people could not identify an oak tree (Orr 2017). It is important that people feel connected to nature if they are going to be willing to protect it.

There are many additional barriers to working-class people feeling connected to nature, because of their concentration in cities and difficulties in accessing nature, as Anita describes here:

I've been walking with a friend recently who … also comes from a very working-class background and we go out into nature hiking, walking, naming trees, picking mushrooms and all that quite regularly and we seem to have really been drawn to doing that together partly because it's something that wasn't available to us … there's this stuff that middle-class people are taught. You do the Duke of Edinburgh Award scheme and all that stuff and we both can see how working-class people are excluded from countryside and nature-based stuff … also [they can] *have less time and energy because of the nature of their labour so maybe they don't want to go walking in the countryside because they've done 50 hours putting scaffolding up. They're tired.* (Anita)

This separation from nature has impacted many in the industrialised world but there are these additional dimensions for working-class people. Through the resulting division of nature and society, it has been possible to separate aspirations for social justice and for environmental justice. This situation is mirrored in the almost parallel universes of environmental and human academic enquiry and discourse (Guerrero et al. 2018). The fact that we need to care for the environment in order to care for ourselves has been obscured. In a study of how to integrate the social and ecological aspects of environmentalism, Guerrero et al. (2018) advocate setting up interdisciplinary research teams as one way of addressing this.

If we link class oppression with environmental harm, we can see that the two need to be addressed simultaneously. Conceiving environmentalism as a socio-ecological problem requires spanning disciplines in a way that challenges the disciplinary approach dominant in most areas of contemporary research. Some academics are now doing this. For example, from the field of social policy, Tony Fitzpatrick (2011, 2014) and Ian Gough (Gough 2013, 2017) focus on the connections between climate and change and poverty and suggest policies for equitable environmental transition. From the fields of sociology and psychology, Nora Räthzel and David Uzzell (2012, 2013) focus on workplace or labour issues and workers' rights and environmentalism. They wish to reintroduce societal relations (relations of production, relations of consumption, political and power relations) into discussions of sustainability and environmentalism (Räthzel and Uzzell 2012). Climate justice and energy justice study and activism have also contributed to breaking down this separation in recent years.

The key issue that connects the human and the environmental realm is the issue of our health. Craig Bennett, the current CEO of Friends of the Earth England and Wales, noted: 'A big blind spot around the debate on the environment is around health. It is an area where we haven't really made the connections we should have done over the last few decades. We've got really strong and established environment and health sectors and actually not that much collaboration between them at the moment. That is extraordinary' (2015, np).

There are now signs of this changing, for example, with the evidence presented in the Lancet Commission reports (2017), as discussed in Chap. 1, but there is much work to do in terms of bringing this knowledge to the wider public. Environmental movements and workers' movements could be potent allies if focused on this question of health. As discussed in Chap. 1, public health discourse continually plays down the importance of the environment. The need to maintain profits under capitalism drives risk-taking with regard to human health (Kovel 2002; Magdoff and Foster 2011). It would not be profitable for people to know how industrial products and processes are undermining their health since they might want to cease purchasing such products. The perceived division between the environment and health has been a major block for environmental justice campaigners who struggle to convince others of the impacts of toxic facilities and commodities on their health.

Co-opted Environmental Movement

While environmentalism has achieved much in terms of environmental agreements, institutions, technology and lifestyle changes, it is failing in the sense that most of the ecological crises are still escalating (Dempsey et al. 2011; Steffen 2015; Dauvergne 2016). Hence, a recent report by the Civil Society Reflection Group on the 2030 Agenda for Sustainable Development

(2018) states that progress towards the goals are 'off-track' and that 'three years after its adoption, most governments have failed to turn the vision of the 2030 Agenda into real policies. Even worse, policies in a growing number of countries are moving in the opposite direction' (p. 11).

Hard regulations informed by the precautionary principle are rare compared to exhortations for voluntary compliance to make necessary change. This could be because, on the whole, the environmental movement has not been radical, in the sense of challenging the primary political and economic drivers of unsustainability. The mainstream environmentalism that emerged in the 1960s and 1970s was mostly apolitical, in the sense of being neither left nor right on the political spectrum (Paehlke 1989). Even where environmentalists positioned themselves as left-wing, most did not oppose capitalism and many disapproved of an equal distribution of wealth. They tended to focus on positive lifestyle choices, such as eating organic food, indicating a liberal, individualist approach to environmental transition through consumption choices. This approach may be inadequate for the achievement of sustainability, as well as off-putting for the working-class people who can see that politics and economics are driving this problem. Dave, for example, was not attracted to environmental campaigning because he felt that the profit motive and corporate interests that lie behind environmental problems were not being discussed. He remarked,

> *The reason I haven't joined an environmental group is because I believe, only by fundamentally challenging the root cause of these problems, are you ever going to move towards any way of eradicating them … I think we've got to be looking on these issues in a far more broader way* [than the environmental organisations are currently doing], *you know, both in terms of what the causes are, and in terms of what the realistic remedies are … I'm not putting everything off to the socialist utopia as a solution …* [but] *we have to be realistic, and we have to address the real issues.* (Dave)

Dauvergne (2016) argues that the environmental movement, as it currently stands, is pursuing an 'environmentalism of the rich', focusing on eco-business and eco-consumption. From this perspective, environmental discourse has become marketised, focusing on issues where there can be a profit-making technological or market-based fix. It has also become individualised, advocating lifestyle changes which enable profit to continue whilst reducing individual guilt.

In some cases, these individual lifestyle changes can create more environmental problems than they solve. For example, we are being encouraged to replace our petrol-fuelled car with a new electric one. Yet, electric cars, whilst running on cleaner energy, cause other ecological problems related to their production. Compared to petrol-fuelled cars, electric cars 'exhibit the potential for significant increases in human toxicity, freshwater eco-toxicity, freshwater eutrophication, and metal depletion impacts, largely emanating from the vehicle supply chain' (Hawkins et al. 2013). Dauvergne (2016) argues that the ideas and principles of those profiting from continued consumption feed the contemporary environmental movement, with many environmental organisations relying on corporate funds to survive. Therefore, they focus on cleaner and more efficient production, rather than questioning the necessity and usefulness of the product.

It is not that individual lifestyle changes are unimportant, but it is the disproportionate emphasis on them that may be problematic. Most of the working-class people I interviewed did have a feeling of personal responsibility for solving environmental problems, even if they felt they were not able to make much difference as an individual, as Amy and Julie describe here:

> *We have to take our own responsibility … we've all got to do our bit … We have to think this planet belongs to everybody and we've all got a stake in it.* (Amy)

> *I do try to use not so much plastic … I try to remember my Bag for Life and things like that. It is trying to re-educate myself and people … To be fair, I have probably not done a lot … Apart from doing all your proper recycling bit that is obviously going to help somewhere along the line, I don't know what else you can do.* (Julie)

As discussed earlier, many working-class people feel they have few options. As a result, when environmentalism has taken on the role of moralising about individual people's behaviour, it has been very off-putting for working-class people (see Chap. 5). Environmentalism then appears to be about 'educating' working-class people, rather than challenging the state or corporations. While focusing on what working-class people are doing and consuming, many environmentalists and the general public often do not get around to challenging the companies and governments that drive and allow environmental degradation and environmental classism in the first place.

A recent study has found that 100 companies alone are responsible for 71% of global carbon emissions over the last 30 years, mostly fossil fuel companies (Griffin 2017). Individual choices are important, in part because we are the customers of these companies, but our choices are constantly being undermined by corporate decisions. Moreover, many of the environmentally friendly 'choices' are readily available only to the more financially comfortable. The focus on individual lifestyles tends to side-line collective organising to challenge the corporations that are putting the future and current wellbeing and survival of our species in jeopardy. The interviewees were often critical of this approach, as Jo discusses here:

> The companies could do more and they know it. They should do more. Look at … [company] … vile, just vile. I don't buy any of their things. Boycott them. They could do more. They should do more but yet they won't because they just want to make money. That is what I think. They are just not interested are they? I don't really know what more needs to happen. We have got this 12 year warning which is really a miniscule amount of time so I am not quite sure what else needs to happen before the government, the UN, whoever, the World Health Organisation actually step-up. (Jo)

As Jo points out, national governments and supra-national organisations are part of the problem. The dominant overarching paradigms for government-led environmentalism in the Western world are the 'Green Economy' and 'Green Growth' agendas. These are now being strongly pushed by international organisations, such as UNEP and the OECD, and being implemented at a national level. UNEP (2011) defines a green economy as one that achieves 'improved human well-being and social equity', while significantly reducing environmental risks and ecological scarcities'. This should make it a supportive paradigm for working-class environmentalism. However, a particular market-based construction of this concept has come to dominate social, environmental and development policy and discourse, while redistributive or rights-based alternatives have been marginalised (Cook et al. 2012). This meaning of Green Economy promotes market mechanisms, such as pricing mechanisms, setting costs to encourage desirable, and discourage undesirable, behaviour, without regard for their impact on poor and working-class communities (as described in Chap. 4). Depending on its interpretation, Green Economy policies often elicit very lifestyle-orientated and technical solutions to environmental problems. For example, as discussed in Chap. 4, putting the emphasis on the household recycling of waste allows corporations and governments to shift some of the costs of waste management onto the public and shields them from demands to reduce or reuse packaging or change environmentally damaging production such as planned obsolescence.

Within this paradigm, it is assumed that capitalism is entirely compatible with environmental protection or even beneficial for it (e.g. Gore 2000; Porritt 2005). Calls to change society or to ban harmful activities are considered to be naïve or detrimental to the goal of a better world. The strategy proposed is to 'decouple' economic growth from environmental harm by modifying our personal habits, inventing new technologies, and using economic incentives and market-based mechanisms, such as carbon trading. However, 'risk society' (Beck 1992, 1995, 1999) and 'treadmill of production' (Schnaiberg and Gould 2000) theorists point out that technology and efficiency measures have achieved little and nations are becoming increasingly less sustainable, producing more toxic waste and rapidly depleting resources. Evidence points to greatly increased environmental damage in recent times, such as an exponential increase in the production and use of hazardous and potentially hazardous synthetic chemicals (Lancet Commission 2017).

In most cases the anticipated decoupling is not happening. While metrics suggest that some developed countries have decreased their rate of use of natural resources without reducing economic growth (decoupling), this is considered to be based on inadequate accounting. Wiedmann et al. (2013), using the material footprint (MF), a consumption-based indicator of resource use, found that achievements in decoupling in advanced economies are smaller than reported or are even non-existent. Without adequate mitigating measures, by 2050 growth would seem to drive resource use to several times the safe limits for human life on this planet. Using extreme market mechanisms, such as high carbon pricing, and the most effective technological means would enable some decoupling (Schandl et al. 2016) but some analysts consider this would still go beyond the amount of earth materials we can safely use each year (e.g. Hickel 2018).

Perceived environmental improvements in wealthier nations are often based on using poorer nations as supply depots, manufacturing zones and waste repositories. For example, the UK had become dependent on China for low-cost recycling to meet its environmental targets, sending around two-thirds of plastic waste there. Since China banned the import of low-grade plastic waste in 2017 due to its toxicity, UK recycling has been in crisis (Greenpeace 2017).

Even when there have been technologically based gains in efficiency, there is often a 'rebound effect', whereby efficiency in the use of a particular resource often leads to a greater overall use of that resource (the so-called Jevons Paradox) (Jevons 1866). Market and technological fixes generally tend to solve only isolated problems, sometimes causing greater problems elsewhere (Wiedmann et al. 2013). Even where decoupling occurs in a specific country or context, it does not seem to be occurring to the extent necessary in the time available to avoid overstepping planetary boundaries irreversibly.

Overly focusing on technological solutions can prevent us from initiating necessary social change. Stan Cox argues, 'Ecomodern megacities, LED-powered Caesar salads, robotic servants, gasoline that lets you turn carbon dioxide into carbon dioxide, the blockchain, renewable-energy fantasies, and countless other innovative schemes illustrate how market forces are always far better at producing energy-hungry technologies than they are at finding ways to reduce consumption' (2018, np), and he warns us of the opportunity cost in that the entrepreneurs and engineers behind these are leading us to an ecological dead-end. Low-tech solutions, though possibly less patentable and profit-making, may be more helpful and sustainable (see Bihouix, forthcoming).

Market capitalism is limited in what it can achieve. Green companies can sometimes do no more than present a green façade—using the language of environmentalism without the corresponding action. Where companies do actually adopt cleaner production methods, it is often of no social benefit in the wider sense, since the product itself is toxic or unnecessary, not the process of producing them, for example, 'green' ammunition for combat (Nammo 2012), that is, bullets which do not contain lead. It is difficult for corporations to make socially and ecologically rational decisions as they must act according to the company interests, whether or not this coincides with the interests of the environment or wider society. Therefore, producers' choices are constrained by capitalism.

Mainstream environmentalists' single-issue focus on clean energy, while seemingly challenging the big capitalist fossil fuel companies, frames environmentalism as a problem that can be solved within the current system. If we did manage to achieve 100% clean energy, what would we do with it? Within the current system we would use it to continue to overproduce and overconsume because capitalism demands endless growth. It seems that the global economy must grow at a minimum 3% each year for large companies to make a profit (Binswanger 2015). We need to look at the deeper causes of our problems, as several of my interview respondents remarked. For example, Dave said:

> *Pollution isn't a separate issue to the drive for profit, and the fact that people will take, or big business will take, short-cuts in order to make money ... It is very annoying, and very frustrating, you know, when you consider the effect it has on people's lives and, generally, in terms of the survival of the planet, because I think there are certain question marks now over whether we're going to be around in the, you know, in the next 100 years.* (Dave)

Global material extraction and consumption has grown by 94% since 1980, and is still increasing (Giljum et al. 2014). Clean energy, though important, won't prevent this but rethinking our economic system, which requires growth at any cost, might. Global GDP growth doesn't reduce poverty. While global GDP has grown by 271% since 1990, the number of people living on less than $5/day has increased by more than 370 million (Hickel 2015, p. 2).

The focus on fossil fuels has created the illusion that we can continue with the current system as long as we switch to clean energy, but this does not address the underlying cause of the problem. So much is produced purely for profit, rather than need, and much is not socially useful at all. The UK, for example, is the second largest arms trader in the world, producing weapons for the world which will mostly never be used and, if they are, can only create social and ecological harm. Direct conflict, nuclear weapons, military training and military-produced contaminants have been found to have an overwhelmingly negative effect on ecosystem structure and function (see, for example, Lawrence et al. 2015), as well as, of course, human health and wellbeing (see, for example, Alexis-Martin 2019). One of the working-class women I spoke to in my earlier research (Bell and Sweeting 2013) said this emphatically, and also made a point about the opportunity cost of military spending:

> It is important that we do our bit, but what we do locally is really just a drop in the ocean. It is important, but we've got to think bigger … The Council could provide solar panels to everyone but spending money on wars abroad and Trident missiles means you can't do that. (Paula)

Though it is important to try to reduce our negative impact on the environment at the individual level where possible, the greatest polluters are not our neighbours and friends but big corporations and state-funded military operations. Sanders (2009) argues that, even if every single one of us made all the green lifestyle changes recommended by environmentalists, the world would still be hurtling towards environmental and climate disaster because of the damage that the military alone does to the earth. The military is one of the world's worst polluters, yet hardly gets a mention in most environmental debates. The UK government subsidises the arms trade through marketing and by aiding with the cost of research and development (CAAT 2014). As the UK is the second largest arms exporter in the world, after the US, this should be a major focus of UK environmental campaigns. However, most environmental organisations do not focus on these major issues—the ones that would challenge the corporations and vested interests in society.

Limited Democracies

As we have seen in Chap. 4, environmental decision-making processes are often not open, inclusive or democratic, generally speaking. Working-class people are rarely involved in scrutinising scientific information, setting standards that reflect their needs and experience, or deciding on interventions to tackle environmental problems. Information around environmental problems is not disaggregated according to social class or income groups, so it is difficult for people to know about the inequitable burdens that they are enduring and the impacts of those burdens. Working-class people are also less likely to access information about the link between environmental factors and health and, therefore, to challenge toxic facilities in their communities, as Ange described.

Their [the government's] *answer is, like, "oh well, the level of pollution is normal." You know, if today's* [level of pollution is] *five - it's "normal", if tomorrow goes to ten, well they will make ten "normal"; and the day after it's 15, they will make 15 "normal" …. The same with the pesticides, the normal level that we can eat safely … I don't believe any of it is normal because my great, great, great grandparents never had any pesticide … I think people believe that if the Government allow this, they're safe, okay? The government's lied to us many times, say that things were safe. It's like when they were giving sickness pills to the pregnant woman … and they say the water for drinking is safe. It's not really safe … Because of* [myself having] *the cancer, I done my research … Can they give me a leaflet saying 'this is the risk and this the pollution we're making with this factory'? … It needs to be an honest one. That's what maybe they can do.* (Ange)

There can be a tendency among working-class people to ignore dangers if they do not exactly know what the problem is or what can be done about it, as described earlier. However, there are also more explicit structural barriers to working-class people challenging environmental issues, for example, the cost of taking those who destroy the environment to court which has become increasingly prohibitive. In early 2017, ClientEarth, Friends of the Earth and the RSPB began legal proceedings against the UK Lord Chancellor and Secretary of State for Justice regarding new rules which removed a cap on court costs, making it risky for individuals and campaigners to bring an environmental case in the public interest. They said these 'will make it virtually impossible to bring a public interest case to protect the environment' (ClientEarth 2017, np).

While democracy, in theory, gives all citizens a voice, in capitalist societies, the power to influence is greater for the wealthy and the large corporations. Capitalism undermines democracy because the need for business to maintain sufficient profit is paramount. In particular, in neoliberal variants, deregulated and globalised markets have undermined the decision-making powers of elected governments. The wealthy have more power in society and can act in their own economic interests to subvert local and national democracy through forming powerful lobbies to shape government policies (as described by, for example, Faber 2008; Magdoff and Foster 2011). Faber (2008) refers to the networks of powerful interest groups that undermine environmental progress as the 'polluter-industrial complex'. These are think tanks, policy institutes, research centres, foundations, legal associations, political action committees, public relations firms and so on, who are committed to shaping environmental discourses and policies. Financed by corporate bodies, they have dominated environmental debates in the US and beyond, working for their own interests while appearing to express objective, factual, expert-based opinion. Brulle (2018), for example, describes a 'misinformation campaign' to redirect the public discussion on climate change in the US so that, as the scientific consensus about the human causes and catastrophic effects of climate change grew stronger over the last 30 years, there was a coordinated effort by conservative foundations and fossil fuel corporations to promote uncertainty about the existence and causes of climate change and reduce public concern.

Examples of similar pressures around the world abound. For example, in the UK, the struggle to ban asbestos was thwarted for many years by the asbestos industry. In an all-out campaign of misinformation and concealment, they set up their own research and public relations organisations such as the UK Asbestosis Research Council and the UK Asbestos Information Committee to promote their message that the public is not at risk as proper controls are in place and asbestos disease is rare (McCulloch and Tweedale 2008). Asbestos use was eventually banned in the UK in 1999, but the number of cumulative asbestos-induced deaths in the country to date has already reached about 77,000, with another 55,000 predicted according to some reports (e.g. Howie 2012).

In a more recent example, we can see the possible influence of the chemical companies on environmental regulation apparent in a report from the EU's food safety watchdog, the European Food Safety Authority (Efsa). It recommended that glyphosate, a broad-spectrum systemic herbicide that the WHO has designated as 'probably carcinogenic to humans' (IARC 2015), was safe for continued use. It was found that pages of the renewal assessment report (RAR) published in 2015 are identical to an application submitted by a glyphosate manufacturer. It is alleged that sections of the report had been copied and pasted from Monsanto's own study (Chow 2017). The recent EU decision on a five-year licence renewal of the controversial product was largely informed by the Efsa report. In a *Guardian* report on the topic, Franziska Achterberg from Greenpeace remarked, 'It calls into question the entire EU pesticide approval process. If regulators rely on the industry's evaluation of the science without doing their own assessment, the decision whether pesticides are deemed safe or not is effectively in the industry's hands' (Achterberg 2017, np). Despite this apparent unreliability of data and the 1.3 million people who signed a petition calling for it to be banned, glyphosate was relicensed for use in Europe this year. Dave, my research interviewee, also pointed out the problem of obtaining correct scientific information:

> *I think that's the bigger problem we've got to look at, you know, of who will actually take big business on? Because I'm, unfortunately, going to sound pessimistic, but when I hear all these things of independent enquiries, I have to seriously question you know, how far most of the people on these panels are actually prepared to go. I think another case in point is the pharmaceutical industry, which I think is absolutely running rough-shod. You know, they're challenged on nothing, and yet the money they're making, particularly out of the National Health Service, is phenomenal … [I wonder] if half the stuff they're actually flogging is necessary at all, you know. (Dave)*

Chomsky (1997) argues that even the media, often viewed as relatively free from corporate and state influence in the Western world, largely limits the availability of contesting views and effectively pushes a narrow spectrum of elite opinion as a consequence of its close ties with corporations and the state. When environmental and social problems are reported, it is usually in a way that is disconnected from the wider issues. It is my perception that working-class grassroots campaigns against environmental damage receive little coverage, unless they also have some entertainment value. The power of the mainstream media lies in its ability to give the appearance of democratic consent, whilst sowing the seeds of confusion and apathy among the wider public.

Some of the working-class people I spoke to discussed how the media, the politicians and even the environmental agencies tended to support business in a way which effectively made it harder for them to make their voice heard. They perceived a closing of ranks, within organisations and between them. The consequent undermining of democracy was a major theme throughout all the interviews I carried out, as in the following excerpts:

I see it as business partners because the [names of public sector organisation] *licence these people* [the polluting companies], *so, as far as I'm concerned, they're in business ... These people have committed crimes in public office—they should be charged, they should be in court because, if it were you or I, then we would be and that's what I see is the unfairness of it you know. There's a lot of big money and I mean big money ... somebody's got their fingers in the till.* (Paul)

It was like, you know, this new type of breed of politician that, you know, wears a suit and looks more like a stockbroker ... so, there was no support from the local politicians ... I think there was a kind of realisation that, when the developer and [local authority], *get together, they can more or less do what they want, you know, and they don't have to actually, you know, meet the needs of the local people.* (Jim)

They know the problem of pollution caused by the incinerator and I suspect ... I believe, they're all corrupted. They just make money. They are businesses ... the [local authority] *is corrupted, or somehow have an interest, so maybe they're being, you know, they're being given some money on the side.* (Ange)

Economic interests were seen to be undermining democracy. The people I spoke to tended to think that those making environmental decisions are more interested in their own agendas and pockets than what would be good for working-class people. For that reason, they were not feeling that they could trust the people making decisions on their behalf about the environment, or anything else.

The actions of individuals in power may be thoughtless, selfish or outrightly corrupt, but studies of corruption show that there are often structural factors which foster their behaviour, such as high levels of inequality (e.g. Rose-Ackerman and Palifka 2016). Capitalism limits people's choices, making it harder to do the right thing because profit and money-making has to come first in an economic system that depends on it. Whether decisions are influenced by illicit or legal rewards and interests, doing what is best for society becomes a secondary concern. As Magdoff and Foster (2011) argue, it is because of the political and economic weight of the large corporations that the state has come to play a very limited role in environmental protection, largely restricted to regulating pollution, with little attempt to restrict what is produced.

Some argue, however, that there are varieties of capitalisms or capitalists, some of which may act in ways that are beneficial for the environment and society. For example, Gibson-Graham (2006) argue that we cannot assume that companies are always and primarily interested in maximising profits because there are such a wide range of ownership structures of companies and varieties of social actors who are involved in them. Yet, overall, capitalism entails the basic necessity for a business to make sufficient profit relative to competitors to survive, whatever their internal structure or the intentions of individuals. This imperative can induce harmful actions. When decisions are made based on what will help their business to survive in the short term, business owners may make decisions which collectively harm society and the environment.

Governments in capitalist societies are often interested in supporting businesses to meet this profit-making imperative. The Grenfell deaths mentioned at the beginning of this book may not have happened had not previous and current (at the time of writing) political leaders considered health and safety as a 'burden on business' under a 'Better Regulation' agenda and the 'Red Tape Challenge'. For example, David Cameron pledged to 'Kill off health and safety culture for good' and, in a speech to the Confederation of British Industry, spoke of Equality Impact Assessments in dismissive terms, as part of the 'bureaucratic rubbish' that gets in the way of British business (Cameron 2012). His government had already cut the Health and Safety Executive (HSE) budget by 33% in 2011. Cameron saw no contradiction between all of this and his pledge to become 'the greenest government ever' (Cameron 2010).

Finally

The above explanatory factors for environmental classism have been separated out for the purposes of analysis, but it is clear that the actual situation will be multi-layered, with many factors reinforcing and interacting. However, having covered the main apparent causes of environmental classism, it becomes evident that capitalism creates the more surface-level issues. Therefore, it is no accident or natural outcome that class inequalities and classism exist, and environmental classism is one aspect of this wider problem. Under a capitalist system, the accumulation of wealth among the owning and middle-classes has produced the disempowerment, hunger, malnutrition, environmental poisoning, lack of environmental goods, health problems and general lack of wellbeing that the working-classes experience across the world. With all these simultaneous challenges, it is evident that the free hand of the market is an inadequate tool for distribution. Human and ecological needs are not necessarily met as a by-product of the pursuit of profit. This free hand needs to be guided by a rational but empathic brain, a brain that listens to the whole of humanity and also considers the wider family of species on the planet. The next chapter discusses the steps that we could take to make immediate reforms, as well as the strides that must be embarked on to bring down the pillars of environmental classism.

8

Supporting Working-Class Environmentalism

This chapter discusses the implications for policy, practice and activism of environmental classism and working-class environmentalism. I outline some of the changes that my research suggests are necessary under general headings, each linking to the causes defined in Chap. 7, starting from the micro level and moving on to the macro level. There is not the space to debate each in depth or to present all the evidence pertaining to each, but I hope they will be explored further in future research and activism.

Organisational Change: Participatory Practice

It is difficult to generalise about what actions to take, at this more individual level, to ensure inclusion because the best guide would be to be led by the wishes of the working-class people you wish to involve or work with. But I feel it is necessary to add some pointers, so that people are not left wondering what they can do about the issues raised in this book. I have created the guidelines in the Appendices explaining how I, as a working-class environmentalist, would like to be treated by middle-class environmental activists, policy makers and practitioners. Appendix A

© The Author(s) 2020
K. Bell, *Working-Class Environmentalism*,
https://doi.org/10.1007/978-3-030-29519-6_8

lists a number of general indicators of inclusive participation. Appendix B gives a related list of guidelines for the particular inclusion of working-class people. The first list is based on my previous work on this topic (e.g. Bell 2014). The second list is based on my own preferences, my community work background, relevant literature (e.g. Goodman 2011) and the interview material for this book.

Below are some of the comments made by the working-class people I spoke to about how to make environmental decision-making and activism more inclusive. They stress the importance of making events and meetings fun; listening to people's concerns and starting from where they are at; putting working-class people at the core of what you do; and having paid working-class leaders and advisors:

> *I just think when people get together and you have a good laugh with people, they make it fun, do you know what I mean? It's* [a better way of doing things for us] *... not all meetings ... It's just a different way of getting together and trying to change something.* (Sharon)

> *Sometimes* [when I went out to the working-class areas to campaign for the Green Party], *like, you know, people would say, "I've never voted Green and wouldn't think of voting Green, why are you here?" And I'd say, "Well, we want to hear exactly what you think." ... I mean, and then they would talk about the crossing, how dangerous the crossing was where they'd put it, or, you know, that waste-land, and "why haven't we got the boxes for the, what-do-you-call-it, recycling?", you know, just loads of stuff come up ... and people would respond if you took an interest.* (Jim)

Working-class leadership is key, whether bottom-up or top-down. This will require organisations adapting their internal structure, systems and culture to align with the working-class people they hope to attract. Newman (2016) looked at a case study of an organisation aiming to really work alongside people rather than on behalf of, or for, them (Action Aid 2016). She concluded that, to do this effectively, organisations must focus on 'who the staff are, how they interact, what language is used and how values and culture are experienced, created and recreated' (Newman 2016, p. 185). The working-class people I spoke to said the same. For example, Pete referred to the Bernie Sanders 2016 presidential campaign in the US:

> ... and one of the things they said that they would do next time that they didn't do, was to have Black people at the heart of the campaign strategy, instead of having them as an add-on. It's the same here. You have to have working-class people and ethnic minorities at the core of your campaign. Not just "... let's not use the word 'Chav' or 'N...'". You have to go, "What do you need to be at the heart of our campaign? How are we going to help your communities to be involved in those issues?" ... You can't win without the working-class being at the heart of it. It's pathetic that people think... "We are in charge. We will tell people what to do. We just wait for them to come along". (Pete)

A positive step in this direction is exemplified by the Unite Union who are now creating opportunities for working-class people to become Labour MPs (Midgley 2017). Unite runs a future candidates programme, preparing new, working-class trade unionists to take on positions in the Party and support the Labour Party's bursary scheme to enable more working-class candidates to finance their campaigns. They want to 'fundamentally re-shape what people perceive an MP to look and sound like' (Midgley 2017, np). Although MPs from more privileged backgrounds can and should campaign on working-class issues, if working-class people see others like themselves in an organisation, and especially if they see them leading it, they are more likely to be open to engagement. As Mick described when he tried to build the campaign against the diesel generators, mentioned in Chap. 4, working-class people were more likely to engage with people that they could identify with, even if they were not entirely similar:

> There is a strong problem with anything that has a green reputation when you go into working-class areas ... Our saving grace was a founder member who is a very large tattooed ex-biker—he definitely did not look green. (Mick)

But the existing working-class members of environmental organisations and committees should not be used tokenistically as diversity bait, but rather have their advice followed about all aspects of the work being carried out. A great deal of diversity work focuses on getting different groups involved in projects and institutions which have been designed by elite groups. For example, there is a desire to recruit 'diverse' people onto Boards and committees. The agenda is often not being set by working-class people and members of those Boards and committees may have no real interest in fully engaging with the different worldview that the working-class or other diverse member brings. All organisations have their own cultures, and social pressure, albeit unconscious, fosters conformity with that culture. This means that to end environmental classism, those who are already in environmental policy making and campaigning groups and organisations need to be constantly mindful of the needs of members who bring a different perspective, giving them space and time to express views that may seem unusual and working hard to see how their ideas could be included and actioned, even if that would mean quite radical change.

In terms of properly valuing working-class people and what they bring to an organisation, it is particularly important that they are able to access any paid employment opportunities that arise, as Mel described:

> *The first step is fair and inclusive employment within those* [environmental] *organisations. Most of these charities have completely White and privileged board members. They have completely White and privileged staff members. The only people who may be of colour or from working-class communities would be volunteers … quite often in the charity sector you're expected to intern or volunteer before getting your first paid employment but, if you come from a disadvantaged background, that's a luxury you can't afford … You can't afford to get those first steps on the ladder which will lead you onto paid employment so, by its very nature, it's discriminating against people who can't afford that … It means that only the privileged will ever get into a position of decision making, into a position of designing these interventions and programmes.* (Mel)

She went on to say that she felt it to be exploitative to continually ask people to volunteer and that it did not take into account the different pressures and different realities that working-class people face, adding:

> *I think, if you want inclusion and participation, you have to be willing to pay people and acknowledge that their time is an asset and that community*

insight is an asset. So if you want to engage working-class people, if you want to engage BAME people, you have to acknowledge the value of their time ... You have to pay them ... If you have more engagement you would get more realistically designed programmes and then you would get greater community involvement ... Whether it's a working-class community or a BAME community, they need to be involved in programme design, in programme monitoring and evaluation ... Stop bringing in these external consultants that are paid ridiculous amounts of money to tell the community what to do. Actually use local expertise, recognise that there are people ... you know, BAME academics, there are working-class academics ... who could be your consultant. Also there could be people who are highly skilled in gardening and nature that might not have any academic credentials who could be a consultant. (Mel)

It is also important to have working-class leaders in environmental organisations and policy making arenas so as to reflect the population demographics in a representative way and, importantly, to be seen as such. During my employment as a community development worker, I became aware that there is sometimes a kind of tribalism among working-class people that has evolved for self-protection. It can be difficult for working-class people to go into spaces where they will stand out because of a different accent, mannerism, clothes, tastes and knowledge, especially when they have been told that all of those are inferior, as described in Chap. 2. Tim Jackson (2005) has explicitly linked identity with acceptance of environmentalism, stating:

That environmentalism is a form of social protest is scarcely news. But the converse is also true, and equally important: resistance to pro-environmental messages and behaviours has to be understood, at least partly, in the context of social identities. Just as environmentalists construe themselves in opposition to certain social groups, so too do those who resist pro-environmental behaviour. (Jackson 2005, p. 295)

For example, surveys examining attitudes to recycling consistently find a hard-core of people who are opposed to recycling under any circumstances (e.g. RRF 2002, 2004). Jackson (2005) considers that people may feel this way, not as a result of weighing up costs and benefits, but because they associate recycling with a certain kind of person and they do not feel they are one of those people. It has been suggested that this tendency towards cultural protest may be exacerbated in highly stratified societies such as those in the UK (Halpern et al. 2003). As Chap. 7 described, the working-class people I spoke to were very distrustful of elite groups and, generally, gave good reasons. Therefore, attempts to influence behaviour must address social identity and trust. The use of role models that would appeal to the target group has been suggested as a way to promote pro-environmental and sustainable behaviours (Halpern et al. 2003), although that strategy risks putting off those who don't identify positively with that role model and it could also be considered somewhat manipulative.

Kaplan (2000) proposes participatory problem-solving to encouraging sustainable behaviours. Rather than telling people what they must do or do without, his approach enables people to work out for themselves how various broadly defined goals can be met. Kaplan distinguishes three ways of instigating behaviour change: (1) telling people what to do, (2) asking them what they want to do, and (3) helping people understand the issues and inviting them to explore solutions. Mostly the first is used, sometimes the second, but it is the third that Kaplan proposes will be most likely to lead to change. I don't personally believe behaviour and lifestyle changes should be the focus of these environmental messages, but rather encouragement to join, influence or create environmental campaigning organisations that can work for change at a structural level.

There is a lot we can do at the community and organisation levels to break down environmental classism and build working-class environmentalism. But, ideally, this would not just be up to the good-will of conscious and caring middle-class people. It should be enshrined in legislation. This brings us to the next category of social change required.

Legislative Change: Make Class an Equalities and Human Rights Issue

Legislative change is needed but even improving working-class people's access to justice under the current legal framework would improve the situation of environmental classism. Friends of the Earth Scotland, who have been exceptional in being particularly committed to environmental justice, are now campaigning for a Scottish Environmental Court or Tribunal (ECT) and an Environmental Rights Centre (ERC). They anticipate that this will help people navigate the complexities of the legal and planning system and enable them to exercise their environmental rights more effectively and push for stronger environmental protections (FoES 2017). If the ECT and ERC are developed and rolled out across the UK, it would clearly be a very positive step.

However, there are many aspects of legislative change that would have an even greater impact in terms of reducing environmental classism. The most impactful legislative change, in terms of addressing environmental classism, would be to bring 'social class' within equalities legislation. In the UK there is no recourse to legal redress if a working-class person feels they have been discriminated against on the basis of class. As mentioned in Chap. 2, the UK now has a single Equality Act (2010) which protects people from being discriminated against on the basis of nine characteristics, that is, sex, race, disability, sexual orientation, marital or civil partnership status, pregnancy and maternity, religion, belief or age. Hence, these are known as 'protected characteristics'. Class is not included as a protected characteristic.

This contributes to the invisibility of classism. For example, the Public Sector Equality Duty contained in section 149 of the Equality Act 2010 requires public authorities to have due regard to a number of equality considerations when exercising their functions and to gather relevant data but, because class is not a protected characteristic, there are no obligations in relation to class. Without this monitoring, the problem of class oppression and exclusion may remain hidden. In addition, Equality Impact Assessments may be carried out under the Equality Act. These are assessments that public authorities often carry out prior to implementing a policy, with a view to ascertaining its potential impact on equality. They are not required by law, although they are a way of facilitating and evidencing compliance with the Public Sector Equality Duty. Again, because class is not a 'protected characteristic' there are no Equality Impact Assessments on the basis of class, so policies are often not considered in relation to their impact on working-class people.

The idea of a 'socio-economic duty' on public bodies was proposed as part of the UK government's Equality Act in 2010 but was never implemented. Although not focused specifically on class, but rather on poverty or deprivation, it would clearly help to reduce working-class disadvantage. There have since been some attempts to bring in legislation to address this gap in the Equality Act. The Equality Trust and Just Fair are asking UK MPs to support EDM 591: commencement and enforcement of the socio-economic duty, though just 78 have signed to date. However, recently, the Scottish government has proposed the inclusion of a socio-economic duty such that all of Scotland's public bodies should be legally required to consider how their decisions can reduce poverty and inequality. At the time of writing, a public consultation is underway on how to implement this proposal (Scottish Government 2017).

Making class a 'protected characteristic' could help avoid much of the interpersonal and institutional disrespect and discrimination that working-class people face. There could be targets linked to diversity policies on class as there are with other areas of equalities practice. Some of this can happen without the need for legislative change. Some organisations already have 'diversity' policies (broadly understood as policies which recognise, value and reward human difference) which include class, for example, my own union, the University and College Union. Some also have 'social mobility' policies, referring to 'the ability of individuals from disadvantaged backgrounds to move up in the world' (Crawford et al. 2011, p. 6). Social mobility policies often target working-class people, even when they do not mention class. Some organisations target people living on low incomes for positive action as a way of acknowledging class. In the US, for example, a report on how to ensure green jobs benefit working-class people and other marginal groups asserts, 'The number of high-quality jobs that are created and filled should be maximized and incentivized, with an aim to distribute them proportionally by race, gender and income level' (Liu and Keleher 2009, p. 14). Such positive action policies could be developed in organisations across the board (although using the term 'working-class', rather than 'low income', for reasons given elsewhere). However, this is unlikely to happen because, on the whole, if there is not a legal obligation to consider class, it will be less of a priority and/or it will continue to get left off the agenda altogether. As far as I know there is no equality and diversity training on class in the UK or elsewhere, even though, as this book reveals, much could be taught that would help reduce classism. Such training could be part of the national curriculum, in an educational package alongside other equalities training.

These policies on class, though helpful, are not as progressive as designating class as a legally protected characteristic. Even if class was brought into equalities legislation, it would have to go further than the current legislation as there is one glaring deficiency in all the UK diversity and social mobility policies, as well as the Equality Act 2010: In many cases they aspire only to 'equal opportunities' and do not propose 'equality'. They accept a hierarchical system. To come closer to ending environmental classism, we need to have economic and social equality, not just opportunities but outcomes, the subject of the next section.

Policy Change: Reduce Inequality

Based on the previous chapter's analysis of the causes of environmental classism, the problem could be minimised by drastically reducing inequality. As Chap. 7 discussed, reducing inequality would reduce emotionally driven consumption and immensely improve the quality of life for working-class people. Inequality damages the whole society, not just the least well-off (Pickett and Wilkinson 2010). Studies and reports have linked inequality to a whole range of social problems, including poor health; unhappiness (e.g. Oishi et al. 2011); crime (Gallan 2018); unwanted pregnancy (e.g. Kawachi et al. 1997); less trust in others (e.g. Fiske et al. 2012; Fritsche et al. 2017); mental illness (Royal College of Psychiatrists 2010); drug addiction; obesity; loss of community life; imprisonment; childhood disadvantage; increased personal debt (Wilkinson and Pickett 2009); child abuse (Eckenrode et al. 2014) and bullying (Due et al. 2009). Furthermore, it is recognised that excess wealth does not make people happy (Easterlin et al. 2010) and it seems that people are fatigued by trying to compete, causing them to detach themselves from others. One in four adults in the UK today has been diagnosed with a mental illness, and four million people take antidepressants every year. It is an epidemic driven, it seems, by social isolation, inequality, feelings of alienation and materialism. The Royal College of Psychiatrists (2010) conclude that 'Inequality is a major determinant of mental illness: the greater the level of inequality, the worse the health outcomes. Children from the poorest households have a three-fold greater risk of mental ill health than children from the richest households. Mental illness is consistently associated with deprivation, low income, unemployment, poor education, poorer physical health and increased health-risk behaviour' (p. 18). Recent work by Wilkinson and Pickett (2019), analysing WHO and Lancet data, similarly suggests that income inequality causes a society an overall greater burden of anxiety, social isolation, drug and alcohol dependence, as well as narcissism and self-aggrandisement.

Inequality is not only harmful, it is entirely unnecessary. According to an Oxfam report, the world's 100 richest people earned enough in 2012 to end extreme poverty, worldwide, four times over (Oxfam 2013). However, many continue to accept and even legitimise inequality (Hadler 2005) and, until very recently (with the rise of Corbynism), there has been no strong mass movement demand for redistribution. 'Equal opportunities', rather than equality, has been the familiar mantra in the UK. This is now beginning to shift, with inequality being condemned, not only by radicals, such as the global Occupy movement drawing attention to the 1%, but also from pillars of the establishment, such as the World Economic Forum (2017) and Pope Francis who described inequality to be 'the root of social ills'.

The UK government, via general taxation, has made some marginal efforts over the last few decades to improve the material conditions of working-class people's lives through funding services and initiatives in 'deprived' communities; for example, the Single Regeneration Budget in the 1990s, the New Deal for Communities in the 2000s and the National Strategy for Neighbourhood Renewal following on from that. The development of measures of multiple deprivation for small areas (Indices of Multiple Deprivation) was used in part to provide the evidence base for the allocation of grant programmes. These policies aimed to close the gap in outcomes between the most deprived and other neighbourhoods. Although the schemes mostly failed, they at least kept alive the principle of governments and local governments having a remit to address poverty and inequality.

However, full or greater equality is not something that is yet being widely demanded. We have tended to focus instead on social mobility. In 1997, Tony Blair declared that his Labour government would build a nation in which 'each citizen is valued and has a stake; in which no one is excluded from opportunity and the chance to develop their potential' (Tony Blair, London, 9 Dec 1997). In 2010, then Deputy Prime Minister, Nick Clegg, said that reducing inequality was for 'old progressives' and 'new progressives' should instead focus on 'reducing the barriers to social mobility', mainly through education. Almost 20 years later, Theresa May, has similarly said that 'We will do everything we can to help anybody, whatever your background, to go as far as your talents will take you' (Theresa May, Downing Street, 12 July 2016). Despite their rhetoric, social mobility has generally worsened under their leaderships. The UK now has one of the lowest rates of social mobility in Europe and is below average internationally. Moreover, the rate of mobility within the UK has actually declined. A recent report of the UK Social Mobility Commission assessed government policies on social mobility between 1997 and 2017, covering Labour, Liberal Democrat/Conservative Coalition and Conservative governments. The report concluded that, overall, there has been little progress in terms of social mobility and that it will take decades before the divisions in education and employment narrow. Its chair, Alan Milburn, stated:

> *These are volatile and uncertain times. When more and more people feel like they are losing out, social mobility matters more than ever before. Higher social mobility can be a rallying point to prove that modern capitalist economies like our own are capable of creating better, fairer and more inclusive societies … The old agenda has not delivered enough social progress. New approaches are needed if Britain is to become a fairer and more equal country. It is time for a change.* (Alan Milburn, Chair, Social Mobility Commission 2017, pp. 1–7)

Shortly after that, Milburn, and then the entire Social Mobility Commission, resigned. This could be read as an admission that there is no hope of change under the current Conservative government. Social mobility has not only not worked, but may even reinforce inequality as it assumes that a hierarchical society is inevitable and we all have to learn how to best climb the ladder. It implies that the solution to disadvantage is to do better in the competition rather than end the competition altogether. Social mobility is not working and, even if it did, it can only move around the deckchairs—there will always be the better- and worse-off.

What we need is equality, not social mobility. Social mobility will not solve the problem of environmental classism—it would just allow a few working-class people to become wealthier and more powerful and buy or demand their way out of their environmental problems and the related insults and injuries described in this book. Yes, they might try to smooth the way for other working-class people to follow but this will be a very slow transition that will take generations. Therefore, Labour's 2017 manifesto now emphasises fairness, rather than mobility. Instead of being content to simply focus on getting a few more working-class children into the supposedly top institutions, it is focusing on good comprehensive provision. This could be a step in the right direction but, in the face of the major ecological and social crises outlined in Chap. 1, if everyone is to have their needs met within the limitations of the planet, it requires a significant redistribution of wealth and much greater economic equality. One way to achieve this goal is to take the step that the next section focuses on, joining together in campaigns for social and environmental justice.

Social Movement Change: Radicalise the Environmental Movement

As discussed in Chap. 7, market-based and individual responses, from changing light bulbs to carbon trading and patentable 'techno fixes', such as geo-engineering to modify the weather, have dominated the governmental, media, social movement and public discourses on solutions to environmental problems. These strategies have had a minimal impact overall and have had little, if any, meaning for working-class people, as described in earlier chapters. If we want to broaden the environmental movement, environmentalists need to radicalise their critiques and tactics. There are six key ways in which the environmental movement could best engage working-class people in terms of their message and focus. All involve becoming more radical. These are (1) explicitly connect the quality of the environment with the level of public health; (2) connect environmentalism with everyday life; (3) link the local to the global; (4) focus on the companies and governments as the main culprits of unsustainability; (5) build coalitions between workers and environmentalists; and (6) connect environmentalism with a call to end capitalist exploitation.

Explicitly Connect the Environment and Public Health

Making this connection would tap into the working-class tendency to be very protective of, and willing to make sacrifices for, their families and communities (see Chap. 2). If working-class people knew about the harm that toxic environments are doing to their loved ones, I believe they would join campaigns against it, even if they would not do it to protect themselves. Therefore, information about the connections between environments and public health is vitally important. Companies and public authorities should be obliged to provide easily accessible environmental information. In 2005, the Labour government ratified the United Nations Economic Commission for Europe (UNECE) Aarhus Convention, which grants the public rights in relation to access to information, public participation in environmental decision-making and access to justice on matters concerning the local, national and transboundary environment (UNECE 1999). However, as we have seen in this book, these obligations are not always implemented (see also Bell 2011, 2014). In the UK, the Environment Agency makes some environmental data accessible at a local level through their 'What's in your backyard?' and 'Your right to know' webpages (a public register of environmental permits and licences). Through these pages the public can access maps of local facilities, though it is hard for untrained citizens to interpret the information given and the controversies around safety limits or unequal vulnerabilities (Environment Agency 2017c) and even this is now under threat. In 2018, it was announced that the service would end, although the webpage now announces that the closure has been postponed (Environment Agency 2018). Much more accessibly, in Norway, the Environmental Information Act (EIA) 2003 mandates that Norwegian businesses and public authorities have to hold and disseminate environmental information upon request (under the Aarhus Convention only public authorities have this responsibility) and access to information regarding pollution that is harmful to health is mandatory (MOE Norway 2017). This could be a very valuable policy, if accompanied by independent support for people to interpret the information.

When working-class people have information about the impacts of the environment on their health, they can be very assertive in demanding change. Pellow and Brulle (2005) note the many successes of the BAME and working-class-led US environmental justice movement at a local level in shutting down major incinerators and landfills; preventing polluting operations from being built or expanding; and relocating residents from polluted areas. However, local successes may mean shifting the problems elsewhere unless there has been a strong network of solidarity built and working-class people everywhere are aware of the impact of the environment on their health. Environmental groups could provide information about health impacts and generally support working-class-led environmental campaigns, such as occurred recently with regard to the poisoning of the water supply in Flint, Michigan, in the US.

In the last year, some working-class people have begun taking legal cases against companies and governments for environmentally linked damage to their health. For example, in the US, a school groundsman named Dewayne Lee Johnson took Monsanto to court for harm to his health caused by a commonly used herbicide product, Roundup. The jury ruled that Monsanto had caused his terminal cancer and ordered the agrochemical company to pay damages. The company is appealing the decision but, since this verdict, thousands of other cancer patients and families have also begun to seek justice through the courts in relation to Monsanto and its German owner, Bayer AG, blaming their glyphosate-based herbicides for their diseases (Levin and Gillam 2018). In another pioneering case in the UK, Rosamund Adoo-Kissi-Debrah has recently gone to court over the death of her daughter, Ella, a schoolgirl who died of an asthma attack thought to be linked to illegal levels of air pollution near her south London home. Although there have now been a number of reports about pollution being linked to deaths in general, there has never been a case linking pollution to an individual death. The UK high court has just granted a new inquest into Ella's death (Laville 2019). The implications of the verdict could be enormous.

Centring environmentalism on health concerns and, particularly, the plight of the working-class who carry the greatest burden will likely bring the movement in alignment with the concerns of working-class people to protect their families and communities. Though some mainstream organisations do some work on this now, and this work is increasing with the recent publicity on the health impacts of pollution, it has been a minor part of the larger mainstream environmental agenda. We need to listen to working-class people and give credence to their insight into what is going on within their own bodies and within their communities and workplaces.

Some environmentalists, particularly deep ecologists, might find this approach to be too anthropocentric as discussed earlier in this book. A more eco-centric position, focusing on ecological justice, can be supported by working-class people, as many working-class people care about nature as discussed in earlier chapters. However, they are more likely to be willing to join struggles to protect nature if they feel allied to environmentalism as a core value. I believe this is more likely to occur when they see environmentalists reaching out to them on their own terms and helping to protect them from the environmental injustices they are experiencing now, the topic of the next section.

Connect Environmentalism with Everyday Life

Social movement literature indicates that organisations for change tend to gain support when people see how an issue affects their everyday lives or the lives of those they care about (e.g. McAdam et al. 1996). People, in general, are more likely to get involved in a campaign when they feel a combination of anger (about the issue) and hope (that they can achieve change). As part of this, environmentalists have to counter the notion that environmental protection is a privilege the working-classes cannot afford. Working-class people, who may be very preoccupied with how to survive today, will find time for environmentalism if it is also connected to their self-identified wellbeing or their other core values. Community development activities and trade unions were formerly the two key ways of organising with working-class people to achieve the change they identified as important and linking causes through solidarity campaigns. However, much of this structure has now gone, as Pete mentioned:

> *Thatcher very successfully weakened the working-class movement, by breaking the power of the trade unions … Society is becoming less caring where, before, it was the working-class that challenged all that, saying 'Oi, what about us?'* (Pete)

Successive Conservative governments have also instigated cuts to much of the community development work that used to take place on estates around the country. Community development had a wide range of goals but always included empowering people to create change, including preparing people to be available for political mobilisation and movement activism (Minnite and Fox Piven 2016). As a former community development worker, I have observed how most of these professional posts have now been lost. We were trained to empower people as agents of social change through undertaking enquiry, problem-solving and critical reflection on situations of real relevance to them. But, with the 'Big Society' introduced by David Cameron, it was assumed that anybody could do community development work, without training, supervision or pay. Much of what then passed for community development became a way for the state to save money. Many of the people I interviewed for this research lamented the loss of lively community spaces and hoped to revive them, as Amy discusses here:

> *I want to see some of the lovely old buildings that have been boarded up in this town being used. It would be expensive, I know, but I would like to see them used for community projects, places where people can learn new skills, where they can meet together, share time and I think that's the way probably most high streets should go now.* (Amy)

Working-class people find ways to support each other, whether the state supports them in this or not. It is just more difficult without external help. Some of the working-class people I spoke to were very positive about the possibility for mobilising communities, even now. Mick, for example, stated:

> ... *working-class communities have always talked to each other, now they can do it digitally as well. Anyone can be taught to Google. Everyone can learn to put that together with their inbuilt 'bullshit' detector ... and then you find you've got more in common with your Sikh neighbour than the* [elites] *... and once you've learnt that you cannot unlearn it.* (Mick)

It is also a positive development that community unionism is now becoming more common, enabling trade unions to organise in communities and among vulnerable sections of the working-class (TUC 2010). For example, the Unite Union in the UK organises among voluntary sector, faith-based groups, students, retired people and those in irregular and precarious unemployment. This is becoming increasingly important because unions have limited power to communicate with the working-class as a whole since there is an increasingly large informal and hidden unemployed labour sector.

Many of the typical environmental discourses fail to engage working-class people because they either require a lot of effort, time or money or seem too trivial to make a difference. At the same time, environmentalists sometimes seem to have little to say about self-identified working-class problems and give little in the way of solidarity or support. For example, working-class people are often engaged in providing informal communal and support services for each other. Although it might be more effective, in the long term, to struggle against the system that is creating these problems, rather than find better ways of coping with it, some of this caring work is literally life or death. If environmentalists became active in supporting working-class people on their current priorities, those being supported might then have the time, energy and inclination to get involved in campaigning on other issues. This should not be done in a manipulative way, treating the working-class agenda as of less consequence, whilst waiting for people to be ready to discuss and think about middle-class environmental issues. It will happen naturally, as people get to know each other and discuss their mutual concerns, as long as respect is shown.

Connect the Local to the Global

Connections between the global and local need to be better understood and considered. As Miller (1995, p. 144) pointed out: 'When there is a separation of producer from consumer, employee from employer, and investor from investment, the decision-making process is freer to ignore the externalities that might otherwise be factored in'. Humans have a great propensity for empathy and, with greater information about other parts of the world now available through the internet, travel and immigration, it is easier than ever to empathise globally (see Rifkin 2010). Injustices can no longer so easily be rendered invisible through distance. Unions and communities need to consider the global impacts of their local actions; otherwise, we can end up 'outsourcing' our pollution and other environmental problems, that is, solving one problem by creating another elsewhere. It is important that workers and communities cannot be played off against each other around the world as countries compete in a race to the bottom by lowering their environmental standards to attract business. Currently, for example, there is insufficient mutually solidaristic interaction between the unions of the global North and those of the global South. Räthzel and Uzzell (2012) argue that when Northern unions practice solidarity in helping Southern unions with their resources and knowledge, it often comes at the cost of these Northern unions influencing the practices of their Southern colleagues. It is as important to be aware of the power dynamics that can occur in working internationally as those that occur on a class basis within nations.

Focus on the Companies and Governments as the Main Culprits of Unsustainability

Chapter 7 discussed how the dominant environmental discourse has become very individualistic, as if personal behaviour change is all that is required. This masks the bigger environmental harms such as unsustainable agriculture, transport, housing and defence policies; and encourages an overemphasis on individual action, at the expense of systemic or regulatory action. Environmentalists need to be prepared to challenge all that is harmful about our societies, including how we are manipulated into supporting wars that are destroying both ecosystems and human health, and start putting this at the forefront of environmentalism (see Chap. 7). Individuals can be highly manipulated by companies and states, so putting all the emphasis on individuals resisting this pressure, such as occurs with a focus on asking people to consume less, would be a slow and difficult process.

Gorelick (2017) describes how overconsumption has been induced in the population over time to support the growth of capitalism. He describes how advertising has been the means used to scale up consumption. Longstanding working-class values such as thrift and avoiding waste were swept away as conspicuous consumption took off, with goods being replaced for the sake of fashion and status, rather than because they were worn out. Advertising has encouraged people to feel inadequate and insecure if they don't have the latest products or fashionable clothes. Those most likely to be vulnerable to this manipulation are those who feel the most insecure—those on low incomes—but also those who have high expectations placed upon them—the middle-classes. Even children now fear being ostracised if they wear clothes or shoes that are out of fashion or cheap (Elliot and Leonard 2006). Consumer culture causes psychological problems by encouraging competitiveness, overconsumption and indebtedness, contributing to an epidemic of depression, anxiety and addictions around the world (Dittmar 2007).

Even if we are able to see beyond the manipulation and try to make our possessions last as long as possible, we may not be able to because of the practices and policies of companies and governments. Repairs are often more expensive than buying an item new, resulting in repairable goods going to landfills. Even if we can afford to make the repairs, and we can find someone willing to do the work, as I have often experienced, there is often no company prepared to sell the parts.

Some enterprising environmentalists have tried to organise ways of encouraging reuse and repair through a network of not-for-profit 'repair cafés'. The first was set up in Amsterdam in 2009 and there are now more than 1300 around the world. They have tools and materials to help people repair their own clothes, furniture, electrical appliances, bicycles, crockery, toy and so on, and there are trained volunteers available to help. However, even these valiant attempts to introduce sustainability within capitalism struggle to survive. One of the Bristol businesses linked to the repair café network is due to close at the end of this year because it could not make sufficient profit without compromising its principles (Roll for the Soul 2017). To reduce the waste generated, we could legislate for extended warranties on products, ban planned obsolescence, and introduce a 'right to repair' so that parts, information and agreements make it possible to repair goods without voiding the product's warranty. In the US, Right to Repair legislation is being passed in many states and there is now a Right to Repair national campaign (see ifixit.org). Short-life goods create profit, though, and companies are simply trying to survive within capitalism.

Build Coalitions Between Workers' Organisations and Environmental Organisations

Unlike mainstream environmentalists, eco-socialists see the working-class as the main social agency that will end capitalist exploitation of humans and nature. John Bellamy Foster, for example, identifies a new 'environmental working-class' that is forging alliances with other oppressed groups in response to environmental degradation (Foster 2013). To support this environmental working-class, we will need to move away from the classic jobs versus environment divisions. Most of the interviewees for this book were not impressed by the jobs argument for keeping toxic and environmentally harmful industries. Julie quipped, for example,

> *People need the jobs in the toxic factories? Well, it's not a lot of good if you're killing them off! ... They might be providing jobs but if they are killing people in the meantime then it is not worth it. Even if they could just regulate their output like it should be. People would still keep their jobs but then they [the toxic companies] would not make so much profit.* (Julie)

Faber (2008) points out how, in the US, the 'jobs versus environment' dichotomy keeps the two most powerful social movements in the country—the social and environmental—separated. Ignoring the role of the working-class in the environmental movement is to lose the opportunity for a meaningful attack on capitalism's exploitation of labour and the environment. It would be more useful to link up trade union concerns for meaningful and secure work with democratic control of the workplace and environmental and human health protection. The concept of 'Just Transition', as discussed in Chap. 6, has been a useful rallying concept in this regard. Crucially, unions could be supported by environmentalists in their role of identifying and campaigning against toxics and other health and safety issues in the workplace. In recent years, an exciting new area of research has opened up in environmental labour studies, led by scholars such as Nora Räthzel and David Uzzell (e.g. 2013), which can support this.

Taking this a stage further, environmentalists and trade unions could work together on eliminating the production of environmentally and socially damaging goods altogether. Trade unions that represent sectors that may be negatively affected by environmental regulations and the transition to sustainability, such as those in the chemicals, mining and manufacturing sectors, are understandably concerned that their members' jobs may disappear in the process. They, therefore, need to be involved in all the discussions regarding this transition, so that everyone who has the capacity can have socially useful work to do in high-status occupations where they are paid the same or better than they are now. Jack, who works in the aircraft industry, also made the point about being involved in the discussions about transition in the workplace, stating:

> I would not like to just be told my job was ending like that, even if they said they would train me to do something else that was more green. I would like to be asked how to solve the problem … I don't think it would work if you just tell people they have been doing everything wrong and now you are going to tell them what to do instead. We just would not believe it because we would feel crap, basically, … humiliated. (Jack)

Such comments, and the wider discussions on just transition, indicate that what is required is an employment strategy based on a worker-led programme which explores ways in which industries and services can become socially useful and environmentally sound. There is the emblematic historical example of the Lucas Aerospace Shop Stewards' attempt to involve workers in the transformation of production from largely weapons into the production of socially useful products (Räthzel et al. 2010). Research by the UK Campaign Against the Arms Trade has shown how a move towards offshore wind and marine energy could produce more jobs than the entire arms industry (CAAT 2014). These jobs would provide highly skilled alternative employment for military industry workers. Renewables jobs have a similar breakdown across broad categories of skill levels and employ many of the same branches of engineering. The newly converted jobs could also support tens of thousands of supply chain jobs that could be located anywhere in the country.

Another interesting recent development is discussion of the idea of Workers on Boards (WOBs). O'Shea (2017) emphasises that WOBs should be members of a trade union, so that they are not just used as part of a public relations exercise. It is also important that they are listened to; that there is more than one worker on each board, so they can support each other; and that they are provided with adequate training and resources.

Strategies such as the 'One Million Climate Jobs Campaign' in the UK (Campaign Against Climate Change 2010) have also been useful for creating links with trade unionists. Other successes are mentioned in Chap. 6, which highlights the various ways unions have joined with environmentalists on the issue of climate change. One particular mobilisation that I was involved in, in 2009, also shows the potential for environmentalists to work with unions. A group of workers in a wind turbine factory on the Isle of Wight occupied the building following an announcement by their employer, Vestas, that the plant would close. Environmentalists and trade unionists from across the country came to support them with food, money and presence. The occupation lasted only 18 days, but rallies took place all around the country and even internationally. The National Union of Rail, Maritime and Transport Workers (RMT) union gave practical, legal and political support and the Campaign Against Climate Change was particularly active (see also Hampton 2015).

It is relatively easy for environmentalists to engage with 'green' sectors where there are 'win-win' scenarios, compared to engaging with workers in those sectors of employment that eventually need to be abandoned without substitution because they are too environmentally harmful and/or not socially useful. For example, Henriksson (2011) suggests converting the car industry into one that produces useful and ecologically harmonious products. He explains that unions need to lead this process since it is the workers that have the technical knowledge regarding how to realise such a conversion. However, if this occurred, what kind of alternative products would the workers feel most competent to produce? Would they advocate the manufacture of electric cars? This is not the most ecologically sustainable transport option (see Chap. 7) but such a transition would enable jobs and profit to continue in that sector. However, overall, in a sustainable society, the number of cars would have to reduce drastically. We may simply not need as much manufacturing, in general, if we organise our society more sustainably. It may be difficult for the workers in the sectors that need to contract to accept this, as much pride is wrapped up in people's jobs as discussed in Chap. 5, and as Jack describes above. It is unlikely that the unions would accept abandoning their specific sector, given their organisation into industries and workplaces. This would particularly be the case at the local level. Räthzel and Uzzell (2012, p. 89) highlight the difference between international and national confederations and federations, noting that while 'an international union may have the vision and resources to develop a broader perspective for environmental policies, a local union will be much more tied to the immediate, everyday interests of their members'. Usually the primary interest will be keeping their job.

An example of a union that has moved beyond the narrow focus on maintaining unnecessary existing jobs is FILLEA-CGIL, the biggest Italian construction union, which is calling for the end of construction of unnecessary new buildings. They state, 'At stake, is not just the future of work in our sector, but also the future of the countryside … we have to stop building for building's sake … we need a new urban strategy, capable of drastically reducing the consumption of land and the use of cement' (in Clarke et al. 2019, p.1). They call for the reuse and restoration of existing buildings and for legislation to prevent the extent of speculation and using housing as an investment. In addition, they call for the use of cement to be reduced by 50% by 2020 and for it to be eventually replaced by more sustainable materials, such as hemp and lime (in Clarke et al. 2019). Data gathering is complex but it is currently generally considered that currently cement production is contributing to about 8% of the world's CO_2 emissions (Andrew 2018). Much of this is used for new building while millions of buildings in cities around the world sit empty and, sometimes, unsold. Statistics published by the Ministry of Housing, Communities and Local Government (MHCLG) put the number of empty homes in England in October 2017 at 605,891. Of these, 205,293 were classed as long-term empty properties (empty for longer than six months) (MHCLG 2018). FILLEA-CGIL recognises that refurbishing of existing buildings would be better for the environment as well as provide socially useful green jobs.

However, in general, 'green jobs' discourses are still often based on the idea of high productivity and high consumption. While many environmentalists challenge this, the dominant international and national discourses on sustainability all commit to continued economic growth. The Rio+20 outcome document spoke of the need for 'sustained economic growth' (UNCSD 2012, p. 2) and the 2030 Agenda for Sustainable Development states, 'Sustained, inclusive and sustainable economic growth is essential for prosperity' (SDKP 2017, p. 1). National economic growth might be necessary within some low-income countries where there is still a need to provide basic services, such as sanitation and clean water. However, globally, as Chap. 5 discussed, we need to reduce growth.

Growth has generally proved inefficient as a means to meeting needs or improving wellbeing and has often contributed to increasing inequality (Harvey 2006; Woodward and Simms 2006; Melamed and Hartwig 2011). Gross domestic product (GDP), a key measure of growth, is of limited use for measuring social progress. With GDP measurement, even a natural disaster would be a gift to the economy, because of the additional economic activity generated by repairs (Stiglitz et al. 2010). Growth does not even reliably deliver happiness, according to a number of studies. For example, Easterlin (2013), studying subjective feelings of wellbeing over time, globally, found that, in the longer term, whether countries are rich, poor or somewhere in between 'there is no evidence that a higher growth rate increases the rate of improvement in life satisfaction; rather, the evidence is that it has no significant effect at all' (Easterlin 2013, p. 4). Family, work, community, friends, health, personal freedom or autonomy, moral values and a sense of meaning or purpose in life have been shown to be more important for wellbeing than material satisfaction beyond a basic level of income (see Jackson 2006; Feeney and Collins 2015). There has been a growing interest in shifting away from a narrow focus on GDP to a more general focus on sustainability and wellbeing. The United Nations Sustainable Development Goals could be seen as a step in this direction, with goals to reduce inequality and eliminate hunger and poverty. There is also now a global Wellbeing Economy Alliance (WE-All) that is coordinating and catalysing the creation of sustainable wellbeing economies (Costanza et al. 2018).

Some environmentalists consider that we should accept that jobs will go in a planned shrinking of the economy and so we need to build a post-work society. Hoffman (2017), for example, argues that modern-day work is inherently unsustainable. She reminds us that the idea of work as a purposeful activity that everyone should spend most of their adult life diligently engaged in, which we today assume as 'natural' and universal, is historically a relatively recent phenomenon. Hoffman (2017, p. 27) argues for 'the opening of public, political debates about the spirit and purpose of work'. Questions that need to be considered include: Which work is, for individuals, biosphere, and society as a whole, valuable and meaningful? Which work is harmful or pointless? How should the remaining work be organised to be socio-ecologically sustainable? Which needs are genuine and to be satisfied in which way, and when does a society have 'enough'?

Hoffman (2017) states that in the post-work society, there would no longer be 'unemployment', as such, but as much to do as collectively decided on. More time would be available for participation in politics, public affairs and community organisation. There would also be stronger social bonds.

However, I believe that, even when non-socially useful work has ended, there would still be a great deal of work to do in a sustainable society. We could employ people in necessary jobs that would be environmentally beneficial, such as mending all the material goods that are now destined for landfill, refurbishing all the low-quality housing and having people do jobs that have been taken over by machines and robots. There could be local and regional self-sufficiency; self-production of food, energy, housing and clothing. There could be organic agriculture, more repair and maintenance, more low-productivity 'hands-on' activities, education, care and helping services and so on. There could be more crafts, accompanied by re-skilling the populace. We could end the throwaway economy. Though throwaway might seem to be good for jobs, many jobs are also lost when throwaway replaces the repairing and cleaning of these goods. When I was a young person, my first job (aged 13) was collecting shoes to be re-heeled and returning them after. Now few people get their shoes re-heeled because the shoes tend to fall apart before they have worn down at the heels. There are not only much fewer shoe repair businesses but also clock menders and electronic and household repairers.

Yet, counter to the throwaway trend, in Minnesota, US, an organisation called Tech Dump trains people who face barriers to employment on how to refurbish and salvage unwanted electronics. Until recently it was difficult for them to do this work because the companies that produce the electronics go to great lengths to make their products difficult or impossible to repair. They do not allow people to access the necessary manuals, tools and parts. However, the State of Minnesota in the US is now considering a bill that would require electronics manufacturers to make available these replacement parts and tools (Schaefer and LaGrange 2018). This would be an excellent policy to enable sustainable jobs.

Available jobs and necessary work need to be more closely aligned. Tim Jackson (2017) notes a 'curious paradox' in that

> *… in the worst centres of unemployment in the world, there is still masses of work to be done. Everywhere I went I saw the need for human labour. Sick people not getting cured. Kids neglected in classrooms. Buildings in need of renovation. A rising tide of mental illness. The encroaching sea of underinvestment. It wasn't the demand for work that had gone missing, but the institutions (and economics) to deliver it.* (Jackson 2017, np)

All manner of jobs are needed but do not exist because they are not profitable. For example, a third of disabled people in England aren't now getting enough support with their basic needs like washing, dressing or eating because of insufficient funding to employ workers (Brawn et al. 2013). All these jobs would probably require government subsidies. This is not impossible. In the UK, we already subsidise every job in the arms industry (Jackson 2011; SIPRI and CAAT 2016). As Tony Benn, former Labour MP, now deceased, often said, 'if we can find the money to kill people, we can find money to help people' (e.g. Benn 2007). A background paper linked to the UN's 2019 Global Sustainable Development Report recently proposed that governments enact a jobs guarantee scheme which would ensure that all people capable and willing to work would be able to get a permanent, state-funded and locally administered job. They suggest that these jobs could be modelled to serve the transition to sustainability and to build capacities to adapt to climate change. They state that 'the job guarantee would ensure full employment. It would lessen insecurity and the need to compete for environmentally destructive jobs on the individual and the collective level' (Järvensivu et al. 2018). There would clearly be plenty of work to do in a sustainable society and such a scheme would help to ensure they are done but we would still probably work less, or stop overworking because the unnecessary jobs could be abandoned.

Would the trade unions accept this? I believe this could happen if workers and environmentalists could join together on a major threat to jobs—the technology that is now replacing jobs purely for the purposes of making more profit. In the past, workers displaced by technology have generally found other jobs but this seems unlikely to continue with the accelerated pace of technological change. Some analysts still argue that AI, robots, driverless vehicles and related technologies should create as many jobs as they displace due to a rise in incomes and as new products are developed (e.g. Price Waterhouse Cooper 2018). However, this view is now increasingly being challenged with, for example, the governor of the Bank of England, Mark Carney, predicting huge job losses as a result of greater automation (Drury 2018). In general, the realisation is gradually dawning that new technology is going to destroy a lot more jobs than it creates (e.g. Varoufakis 2017; Lowy Institute 2018), with predictions of up to 80% job loss in some sectors (Frey et al. 2017). There appears to be no plan for replacing these jobs. Environmentalists should be arguing for the protection and/or replacement of jobs that are now being lost to machines and robots. Technology which destroys jobs, even while having some other advantages, is not socially progressive. Though we are told it would eliminate mundane work, work is often made boring by the way it is organised, and can be made even more boring when all there is to do is check a machine is working (ask the staff who watch over the self-serve tills in supermarkets). Mark Carney has recently described how, with greater automation, work becomes less effortful but more tedious and tiring (Drury 2018). He also predicted a rise in inequalities between the high-skilled workers who benefit from the new technology and those who are side-lined by them. This analysis was also recently echoed by the Lowy Institute (2018) which claimed that the future high-tech world will be one of far greater inequality.

All useful jobs must be protected in the transition to sustainability. However, realistically, the post-growth society, if it is not to result in wide-scale poverty and job loss, also requires taking the next step.

Connect Environmentalism with a Call to End Capitalist Exploitation

This is the most radical step of all, but it is really the only way that the new environmental discourse would make sense (to avoid repetition, this will be discussed in the last section of this chapter).

Political Change: Create an Eco-social State Based on Sharing

If working-class people are to get on board with an environmental movement that is ultimately requiring them to transform their work, it needs to be in a context whereby people can feel safe to do this, that is, not within the current neoliberal, seemingly uncaring, state. If we reduce growth in such a situation, even for sound ecological reasons, we are basically talking about a green austerity programme. Austerity policies have devastated the lives of millions of working-class people across Europe since they began to be imposed from 2008 (Blyth 2013), so that could never be part of the solution. We have also seen throughout this book that discourses of 'sacrifice' and threats to jobs are not attractive to working-class people. Even the idea of green jobs that may come later is not as attractive as keeping the job you have now (Räthzel and Uzzell 2013). We need to offer working-class people a way to embrace the measures necessary without fearing unemployment, underemployment and a drop in income.

There is no necessity for people to go without jobs, services or essential goods. The funds to pay for these exist but are no longer under the control of the governments that could use them for social development. As the World Inequality Report 2018 states, 'Over the past decades, countries have become richer, but governments have become poor' (p. 14) due to a massive shift towards private capital. To encourage policies that would put this wealth to use for humanity and the planet, social movements could campaign for state actions to redistribute wealth. Yet, Peter Newell (2005) has argued that the middle-class membership base of some environmental groups results in their reluctance to campaign on issues which challenge property relations or access to land. He stated, 'The political interests of the class they represent make it unlikely that anything other than a weak sustainability reformist agenda can be pursued given the extent to which they benefit, directly and indirectly, from the status quo' (p. 84).

This may be the case for many but, given the urgency of the situation, and the recent publicity on how all will benefit from a more equal society, perhaps some would now be willing to accept and work for change. Reducing inequality seems imperative for sustainability. Ian Gough (2017, p. 179) argues, 'When the cake shrinks, its distribution becomes critical'. Like environmental transition, this redistributive transition will require political responses to create long-term societal change.

Working-class organisations, themselves, can also mobilise to press for change. Recent studies show that, around the world, unionisation appears to have a significant equalising effect on national income distribution, with less inequality across society where union membership is higher (IMF 2015; Farber et al. 2018; IPPR 2018). This is mainly because wages and jobs are protected.

The degree of redistribution necessary will also require governmental action and the transformation to, what has been variously referred to as an ecological state, an environmental state or an eco-social state (see Jakobsson et al. 2017). So, rather than talk about 'sacrifice' and job losses, we could be making the case for a state-led redistribution of wealth, as a means to meet human needs and wellbeing within the limits of the planet. I believe this would be a campaign that many working-class people would support. We could also call it 'sharing', although, unfortunately, the term 'sharing economy' is now being used to mean market models of peer-to-peer exchange in which people rent beds, cars, boats and other assets directly from each other. This economy has its advantages but, largely untaxed and unregulated, it is probably never going to seriously redistribute wealth.

Redistribution would require the eco-social state developing radical social policies that would make life better for everyone. Redistribution would mean that, while we might be 'poorer' in terms of individual material goods, we will be richer in terms of health, public goods and human relationships. Some social policy academics and other analysts have been working on what this might look like. For example, Kate Raworth (2017) uses a doughnut-shaped diagram to characterise a new sustainable economy. The outer circle of the doughnut represents the productive limits of the earth's systems, the ceiling that should not be exceeded. The inner circle represents the foundation for human survival and wellbeing. In between the two circles is the 'regenerative and distributive' economy.

Such policies are based on the belief that, instead of seeking to maximise growth, we should set a lower threshold of wellbeing, below which no one should fall, and an upper threshold of environmental limits that economic life should not transgress. Work is already taking place on what the lower threshold should look like in different countries. For example, since 2008, the Minimum Income Study in the UK took focus groups and asked the participants what they thought households needed to reach an acceptable minimum standard of living (Davis et al. 2016). If people had an adequate minimum income or a 'decent life budget' to live on, overall carbon emissions in the UK would drop by 37% (Druckman and Jackson 2009). However, to reduce our consumption to 6%, as Fisher (2013) estimated that affluent countries need to do to prevent irreversibly overstepping planetary boundaries, we would still need to do more (see Chap. 5).

Some argue for a progressive consumption tax. David Fell (2016), for example, makes the case for a 'smart VAT' with higher rates for harmful consumption in terms of health and wellbeing or environmental damage.

However, as discussed in Chap. 4, such market mechanisms tend to have negative impacts on working-class people and they could even elevate the status of the higher taxed goods as they come to be seen as the prerogative of the wealthy. It would be better, I believe, to move towards a cap on excessive incomes and wealth. Ian Gough notes, 'The only sure ways to provide security of income in a post-growth society will be to tax or socialise a substantial part of private wealth' (2017, p. 181). There could also be a maximum income in each country, for example, as a 100% income tax above a certain threshold, to combine the income 'floor' with a 'ceiling'. This would contribute to ending socially and ecologically unjust and harmful extreme wealth (Schmelzer and Passadakis 2011; Alexander 2015). The OECD (2018b) have recently suggested that governments should consider deploying the taxation system to reduce wealth inequality, with inheritance tax the favoured method.

Gough (2017) also advocates widening social consumption as part of an eco-social strategy. He notes that this would reduce opportunities for people to compare consumption, one of the drivers of excess (see Chap. 7). A state which enables greater social consumption, as opposed to private consumption, would reduce inequalities spectacularly. Public services make an enormous contribution to income and reduce income inequality in OECD countries by 20%, on average (Verbist et al. 2012; Seery 2014). Gough (2017) also notes that public consumption is more ecologically efficient than private consumption, with publicly funded welfare states emitting less carbon than countries which rely on private alternatives, in part as a result of better allocation of resources.

Monbiot proposes the idea of 'private sufficiency and public luxury'. Since the environment cannot sustain everyone enjoying private luxury, we could all have access to free-at-the-point-of-use swimming pools, parks, playgrounds, sports centres, galleries, allotments and public transport. He suggests that funding should be allocated by participatory budgeting as in the Brazilian city of Porto Alegre, with citizens deciding on how money should be spent. Monbiot states, 'The results—better water, sanitation, health, schools and nurseries—have been so spectacular that large numbers of people now lobby the city council to raise their taxes. When you control the budget, you can see the point of public investment' (Monbiot 2017a, np). A similar suggestion for Universal Basic

Services was made recently by the Institute for Global Prosperity (2017). Their recommendations for the UK include building 1.5 million new social housing units, offered for free to those in most need; a food service providing one-third of meals for the 2.2 million households considered to experience food insecurity each year; free bus passes for everyone; basic phone services, the internet and the cost of the BBC licence fee. Some countries and cities are already piloting or developing free public transport, for example, Seoul, Tallin, Estonia and Germany.

The Labour Party is looking at these free basic service proposals as a programme they might offer when in government. Although they frame the necessity for this in terms of the rise of technology (again, assuming its inevitability), this policy would help to equalise society and undercut the need to consume for status reasons, as well as ensuring basic needs are met. Another idea comes from Warwick (2017) who suggests that students should spend a year working the land before university. He argues that a year of 'eco-conscription' between school and university would help people to reconnect with nature and the land. Obviously, we do not all go to university but a year of ecological work for young people would help to get the message across that we are in a state of emergency, as well as enabling some useful work to be done.

There have also been calls for a universal basic income (UBI), a regular payment from the state to every citizen, on ecological grounds, as this would break the necessity to have jobs at any cost. However, a UBI comes with significant advantages and disadvantages. Depending who is in charge of the UBI programme, it could also aggravate inequality and reduce social programmes. Some suggest UBIs are intended to provide political cover for the elimination of social programmes and services (e.g. Kleiner 2016). However, an ecological UBI could accompany more extensive social programmes and services and be linked to ecological work programmes. As discussed earlier, there are so many socially useful jobs that need doing but are either not done or not paid for, because they are not profit-making. A UBI payment could pay people to do those important jobs. It could involve those with the capacity to do so reciprocally working on eco-projects, even if it is just for two hours a day (or one month a year—people could choose the most convenient).

So-called contraction and convergence across nations will need to be part of this vision of sharing, with every country bringing its GHG emissions, per capita, to a level which is equal for all countries, leading to a contraction for some countries, and an overall convergence (Meyer 2000). This would involve first going through a phase of planned and targeted economic contraction (not austerity, which cuts the public sector) in the richer countries before stabilising at zero-growth or steady state. In the Global South, economies would grow and redistribute until the basic needs of all were met, when they too would stabilise their growth (e.g. Lawn and Clarke 2010). For this to be valuable for working-class people, the inequalities within each country would also have to be reduced or ended, taking the measures discussed above.

Environmentalists and workers can campaign together for this ecological state. There are growing calls for greater levels of state intervention, including nationalisation of utilities (Legatum Institute and Populus 2017). The Legatum Institute, a think tank, and Populus found levels of support for nationalising large parts of the economy. A strong majority of respondents thought that some of the UK services and utilities should return to public ownership, including water (83% of those polled), electricity (77%), gas (77%) and the railways (76%). There was also some support for nationalising banks and airlines. We could also share out the costs of goods and services. For example, flood risk premiums can be extremely high in areas where flooding is more likely. This market-based and individualised approach to flood insurance contrasts with that established in France and a number of other countries, where disaster risk is socialised, shared amongst all insurance holders with a standard premium applied regardless of location. In France, those people living in a flood risk area pay no more for their home insurance than others, making insurance cover far more affordable and accessible than in the UK.

If it is eventually necessary to contract the economy so as to live within planetary limits, by implementing the above policies, we should ensure everyone has support to lead a dignified, decent and pleasant life. Some of the changes could happen now but it is very unlikely that most of this would happen within the economic system of capitalism. Therefore, we need to think about the most fundamental change, the dismantling of capitalism.

Revolutionary Change: Dismantle Capitalism

As this chapter and most of the book implies, explicitly or implicitly, within capitalist society it is often not possible to make the changes that would be required for sustainability. While global consumer culture and state militarism contribute significantly to overstepping all the planetary boundaries outlined in Chap. 1, as well as ultimately failing to meet human needs, they seem virtually unstoppable within capitalism. This is why I and others have continually argued that the environmental crisis is an inevitable result, not just of neoliberalism, the most extreme variety of capitalism, but of capitalism itself (e.g. Bell 2011, 2012, 2014, 2016). Some of the working-class people I interviewed were also of this opinion. For example, Dave said:

> ... *the only basis on which we will* ... [solve the environmental problems] *is in the context of challenging the system, the profit system itself, and eradicating it, because I can see no solution to the environmental problems otherwise* ... *So, the long-term solution to the problems that face humanity both in Britain and on the world-scale, is the transformation of society on the basis of the ownership of the means of production and planning in the interest of working people.* (Dave)

It is increasingly being recognised that capitalism is not compatible with sustainability and the need to dramatically reduce the consumption of some (e.g. Klein 2014; Monbiot 2017b; Gough 2017). Ian Gough (2017), for example, recently wrote 'to avoid a devastating impact on material living standards, it will be necessary to question and partially dismantle a defining feature of capitalism—the private ownership of the means of production' (p. 181) and finally concludes with, 'In my view post-growth is incompatible with any form of capitalism' (p. 209). It has even been suggested that what we call the Anthropocene, a proposed epoch dating from the commencement of significant human impact on the earth's geology and ecosystems, should really be called the 'Capitalocene' because it is the capitalist system, not humans per se, that has so degraded the planet (Moore 2015).

Some are saying that it is no longer about whether to end capitalism but when and how. For example, Yannis Varoufakis (2017), the ex-Greek Finance Minister who resigned in resistance to the EU imposing austerity on Greece, has recently been on a speaking tour asserting that capitalism is ending, whatever we do, because it has made itself obsolete, in particular through the rise of artificial intelligence. Similarly, Paul Mason (2015) argues that capitalism will not be abolished by protests but by external shocks and changes that have occurred over the past 25 years, including the coming wave of automation. Even the governor of the Bank of England, Mark Carney, has said the automation of millions of jobs could lead to wide-scale unemployment, wage stagnation and the growth of communism (Drury 2018). This may not be the ideal way for capitalism to end since it would plunge the worst-off into unemployment and poverty. Planned reforms could lead to a better ending of capitalism.

If implemented, the above strategies for ending environmental classism and embracing working-class environmentalism would, anyway, potentially transcend the existing capitalist system because they would be a challenge to the current values and structures. There would be a re-evaluation and re-configuration of why, how and for whom goods are produced. Mason (2015) points to the rise of collaborative production outside of the market system—parallel currencies, time banks, cooperatives, new forms of ownership, a different side to the so-called sharing economy—as offering an escape route from capitalism. A state-led first step would be to take resources and services back into public ownership as many now advocate (e.g. We Own It 2017).

However, many environmentalists still consider that environmental harms can be managed within the parameters of reformed capitalism, or even the current neoliberal capitalism. Some are unsure about leaving capitalism behind because they don't like the obvious alternative, socialism. They would argue that socialist countries have a worse environmental record than capitalism and that socialism is based on an environmentally damaging philosophy. Socialism has often focused purely on increasing productivity, rather than aiming to achieve any balance in terms of the rest of nature. However, socialism is a diverse and contested philosophy. Debates abound about the degree to which Marx was pro- or anti-ecological with three main positions—that Marx was deeply anti-ecological, seeing nature as merely a means to achieving human ends (e.g. Victor Ferkiss); that he considered ecology, though eventually chose to take a 'Promethean' (pro-technological, anti-ecological) approach (e.g. Robyn Eckersley); and that he was profoundly ecological and this formed part of his analysis of both capitalism and communism (e.g. John Bellamy Foster). Certainly, the Marxist idea of 'to each according to his/her needs' challenges overproduction, overconsumption and commodity fetishism. Whatever Marx's position, it is evident that twentieth-century socialism succumbed to the same ideology of productivism that was generally prevalent around the world at that time as Foster notes (2008). However, it seems that environmental problems in socialist societies have been exaggerated for ideological reasons. The overall environmental record of former socialist countries may have been no worse than that of capitalist states around the same time (see, for example, Manser 1993; Dominick 1998).

With the new understandings of planetary limits, the record of twenty-first-century socialism appears to be somewhat different. Bolivia's National Development Plan focuses on living well in harmony with nature and also includes specific goals for 'joy' and 'happiness' and 'the construction of a less consumerist and less individualistic society' (MPD 2016). It is also delivering on all the Sustainable Development Goals and its environmental legislation gives rights to nature (Bell 2016, 2017). Cuba was rated as the only country in the world to be developing sustainably in terms of meeting human needs within the capacity of the planet (WWF 2006). The country still uses much less energy, produces less CO_2 and has better air quality than the average Latin American country or country of a similar income bracket, and has the highest proportion of protected forest in Latin America and the Caribbean (Bell 2011, 2014). These countries are, of course, constrained by being part of a predominantly capitalist global world but their accomplishments are remarkable for middle-income countries. Socialism may not be the answer everywhere but it does, at least, show that there are alternatives to capitalism. Establishing capitalist exploitation as the common enemy could lead to effective relationships between workers and environmentalists on many campaigns.

In general, whether it is a break with capitalism or a break with any or all of the barriers to building working-class environmentalism, my research for this book and my wider contact with communities clearly indicate that people are ready for change. There is a great deal of anger among working-class people, with many now ready to take action, as the Grenfell protests in London and around the UK and the *gilets jaunes* in Paris have shown. This anger is very apparent in this book with the distrust, frustration with, and condemnation of, elites frequently apparent in the comments made by the interviewees, as in these excerpts:

I think we have got to … do something about it instead of just yap, yap, yap, talking. All these countries are getting together and nothing has been done and everyone says, '12 years is all we've got', but no one listens. Just act, for God's sake! (Phil)

The governments around the world need to do something about all the poverty while we've got billionaires … I cannot understand why this is being allowed … Who ever we elect … we are told by the big multinational corporations, 'no you do what we tell you to do or you will suffer' and that actually is what is happening. (Bob)

Even those who have until now been quite compliant are ready to take action for change. Amy, a woman in her 60s, for example, said:

This is the first time in my life I've been ready to hold up a placard. But I would do it now because I am so pissed off with the politicians. We need normal people in Parliament. Not these lot. They are so devious and underhanded. People are getting really angry now … they can see these people [the MPs] *represent themselves, not us.* (Amy)

This anger could lead to a strong force for working-class environmentalism but it could also result in a backlash against transitioning to sustainability if it is perceived as a programme of the elites and likely to harm working-class people.

Finally

Like many middle-class people, the working-class people I spoke to and have known throughout my life and work almost unanimously love nature, care about the planet and want to live in a healthy environment. However, they are often put off mainstream environmental movements and environmental policies by attitudes, behaviours and structures that exclude them or even threaten their wellbeing and health. Those I interviewed mostly do not work in environmentally harmful industries and so, in that sense, have less reason to feel threatened by a transition to sustainability, but even these people fail to connect with or are angry about some environmental policies and campaigns.

Whilst mainstream environmentalists continue to agonise about how little their messages are taken up by the wider population, there is a simmering working-class environmentalism ready to be supported and followed. Yet working-class environmentalism, rather than being cultivated, is often challenged by divisive discourses. If environmental and social justice is not better aligned, highlighting the links between inequality and environmental destruction, this environmentalism of the working-class could dissipate or even result in a backlash against sustainability.

Just as I am finishing this book, some exciting proposals are being made that do align these issues. In the US, Alexandria Ocasio-Cortez and a number of Democratic colleagues are calling for a 'Green New Deal' as a plan to transition the US economy to becoming carbon-neutral over the next ten years. This is a call for comprehensive legislation for a mass mobilisation of people and resources towards solutions to the climate crisis. It is seen to include green jobs, electrification of transport, retrofitting buildings with insulation and so on. It calls for a just transition for workers and low-income communities, indigenous communities and communities of colour (Baynes 2019). In the UK there have been similar calls from the left for a 'Green New Deal'. In March 2019, the shadow secretary for Business, Energy and Industrial Strategy, Rebecca Long-Bailey, launched a year-long 'call for evidence' to inform Labour's 'green jobs revolution'. However, this might require a revolution within the Labour Party itself, since it would need to influence other aspects of its policy, such as their Industrial Strategy which is very limited in terms of consideration of environmental issues. In recent days, as I write (April 2019), two MPs presented a bill in an attempt to lay the foundations of a UK Green New Deal. If passed, it would force the Conservative UK government to implement a ten-year public investment plan that prioritises decarbonising the UK economy; eradicating inequality and supporting community- and employee-led transition from high-carbon to low- and zero-carbon industry. The 'decarbonisation and economic strategy' is a private members' bill, tabled by Green Party MP Caroline Lucas and Labour MP Clive Lewis.

Critics of these Green New Deals emphasise that the plans do not address the problematic globalised capitalist economy, which prioritises profits above the environment (e.g. Bellamy Foster 2019). However, Bellamy Foster, for example, does recognise the importance of the mass mobilisation that the Green New Deal calls for and the innovative forms of financing in the US model. His concern is that corporations have a tendency to capture the regulatory process and, therefore, he advocates a change in the ownership of the means of production.

Science is pointing to the extinction of human civilisation and so radical action taking us towards social and environmental justice should not be off the table. As the above discussion indicates, the superficial changes necessary can be difficult without the more fundamental changes. However, it makes sense to start with whatever changes are now possible and able to gain public support. There is much that local authorities and other policy makers could do to reduce or eliminate environmental classism, without necessarily setting out to dismantle capitalism—implement the Aarhus convention on access to justice in environmental decision-making; carry out equalities impact assessments on environmental policies in relation to class; advise people of environmental threats to health; create by-laws and strict environmental standards; make all environmental services free at the point of use. Ultimately, though, we need to go deeper.

Where this deep change should start is always controversial. There may be disagreements between those radicals who argue for a cultural revolution—a revolution in values and ideas as the root of all changes—and those who argue for an economic revolution—a change in ownership and control of the means of production as the driver for the cultural revolution.

I believe that we probably need both and, as individuals and groups, we should focus our research, policy and activism on whichever makes sense to us and enables us to passionately continue to struggle for a better world for everyone. We can use whatever power we have to make these changes in all the places that we have influence. This will include making changes within our own minds. Whether we identify as middle-class, working-class or no class, we must learn to treat all people and nature respectfully; resist classism of all kinds; let ourselves be led by working-class and other disadvantaged people wherever appropriate; and appreciate the importance of the actual and potential contribution of working-class people to environmental transitions.

It will be an ongoing project to rid ourselves of our class-based sociali-sation but we cannot follow the maxim 'change yourself first' because we will never get around to doing anything else. It has to begin with our-selves, but, beyond that, we must also build alliances between social and environmental movements and extend those movements to support the widest scope of humans and ecologies possible. This alliance could help us to build a more equitable, fair and sustainable society. Working-class people and their supporters will need to demand change. The working-class people who gave their time to tell me about their experiences of environmental classism and working-class environmentalism for this book, and for my other projects, want to see this change. In the spirit of amplifying voices, as I set out to do, I give the last word to Paul, who, when I asked if he had anything else to say, closed with:

I don't know what your research will output but I would hope it might be a size eight boot up the ... relevant rears. (Paul)

Appendix A: Checklist for Meaningful Inclusion in Environmental Decision-Making (General)

- All parties that were affected by environmental decisions were invited to contribute to the decision-making process;
- The relevant rules and procedures were applied consistently, with regard to different people and at different times;
- Those affected received accurate and accessible information—that is, timely, honest, easy to understand, digestible and easily available;
- A fair outcome resulted from the process, in terms of substantial and distributional environmental justice;
- There was authentic, accessible and honest communication;
- All parties were accountable—that is, responsible to answer for their actions and decisions and to remedy them if necessary;
- All parties would have access to sufficient material resources to enable them to participate on an equal footing;
- Those affected were included in all stages of decision-making;
- Sufficient skills and personal resources have been available for those affected to participate on an equal basis;
- All participants in the environmental decision-making process were treated with equal respect and value;
- All environmental decisions were made publicly;

© The Author(s) 2020
K. Bell, *Working-Class Environmentalism*,
https://doi.org/10.1007/978-3-030-29519-6

- The environmental decision-making process was open to all questions and alternatives;
- All affected had an equal right and an equal chance to express their point of view;
- There was a lack of external coercion;
- Decision-making was deliberative, that is, free from any authority of prior norms or requirements;
- There is freedom of association;
- There is the right to peaceful protest;
- Those affected had control of the outcome of decisions (ideally proportional to how much they would be affected);
- Consensus decision-making was carried out, whenever this was practical;
- Use of a strong precautionary principle;
- Free access to legal redress;
- The above criteria have not been met through undermining the environmental justice of other species, nations and generations
- (also published and developed in Bell 2014).

Appendix B: Guidelines for Working-Class Inclusion

Avoid tokenism which could allow a working-class person to be isolated. If we are the only 'different' voice, it is very likely we will eventually silence ourselves to keep ourselves safe.

In talks, panels, press conferences or any public event, have a balance of people of different classes, just as there should be a balance of gender, ages, races and so on.

Make membership fees and events costs available at different rates with a realistically proportioned difference between the top and the bottom rate (e.g. middle-class people often earn at least twice as much as working-class people do, but concessions are never given at 50% rate).

Practice positive action in hiring staff, where possible (lawful in the UK under the Equalities Act but only for those with protected characteristics). This would mean choosing a working-class person over a middle-class person when they have the same skills, experience and knowledge, and can be justified on the grounds that they have probably had to work much harder to get the same outcomes on their CVs.

Be led by people who are from working-class communities or backgrounds when making decisions or devising policies or programmes.

Ensure working-class people are involved in any decisions that impact on working-class people—'nothing about us without us' as disabled people say.

© The Author(s) 2020
K. Bell, *Working-Class Environmentalism*,
https://doi.org/10.1007/978-3-030-29519-6

Identify and monitor the class backgrounds of staff and decision-makers.

Challenge classism wherever you see it, including in the media.

Pay all staff equal and sufficient wages, and generally pay people to work wherever possible.

Be aware of 'middle-class' cultures in an institution and check if this limits who can participate.

Provide expenses so that working-class people can attend meetings, for travel, childcare and other caring responsibilities.

Make sure meetings and events are welcoming to working-class people and take place in the areas and building that they normally attend.

Remember that working-class people tend to be 'time poor', so do not have unnecessary meetings and do not expect people to read lots of information or engage in online debates.

Use plain language that is accessible to all.

Think through every activity in terms of how it would impact someone who had a low income, a lack of time, and was feeling stressed.

Go to working-class areas and events and work alongside people on the issues that are important to them.

Put on a variety of cultural events that appeal to both middle-class and working-class people (obscure, classical and folk music generally do not appeal to a large section of working-class people).

Don't stereotype working-class people in terms of what they can do—encourage them to write and be spokespeople and leaders, not just leafletters and caterers.

Find out which working-class organisations exist, develop links with them and try to work together on mutual concerns.

Try to look at issues from the point of view of someone who is not well-off financially.

Create a 'safe space' where classism is not tolerated. Don't be judgemental or patronising.

Try to make everything you do pleasant and relaxing. People with difficult lives or lacking in confidence do not want extra hassles.

If you are middle-class, be honest about your privilege and don't try to pretend you do not have it; leverage your privilege to support others; be prepared to do boring and behind-the-scenes work too; don't dominate meetings or movements or think you have all the answers; be grateful for working-class input.

(This last guideline is taken from Vosper (2016) and some of the others are influenced by her article).

References

Abram, S., Murdoch, J., and Marsden, T. (1996) The Social Construction of "Middle England": the Politics of Participation in Forward Planning. Journal of Rural Studies, 12, 335–364.

Abrams, H.K. (2001) A short history of occupational health. Journal of Public Health Policy, 22(1): 34–80.

Acheson, D., et al. (1998) Independent Inquiry into Inequalities in Health Report, The Stationary Office, London.

Achterberg, F. (2017) in Neslen, A. EU report on weedkiller safety copied text from Monsanto study, The Guardian, 15 September 2017.

Action Aid. (2016) Who we Are, Action Aid International. https://www.action-aid.org.uk/.

Agyeman, J. (2001) 'Ethnic Minorities in Britain: Short Change, Systematic Indifference and Sustainable Development', Journal of Environmental Policy and Planning, 3(1): 15–30.

Agyeman, J. (2002) Constructing Environmental (in)Justice: Transatlantic Tales. Environmental Politics, 11(3): 31–53.

Aiken, G. (2012) Community transitions to low carbon futures in the transition town network (TTN). Geography Compass, 6(2): 89–99.

© The Author(s) 2020
K. Bell, *Working-Class Environmentalism*,
https://doi.org/10.1007/978-3-030-29519-6

Alexander, S. (2015) Basic and Maximum Income. In G. D'Alisa, F. Demaria, and G. Kallis (eds.), Degrowth: A Vocabulary for a New Era. London/New York, Routledge.

Alexis-Martin, B. (2019) Disarming Doomsday. London: Pluto Press.

Allen, B.L. (2003) Uneasy Alchemy: Citizens and Experts in Louisiana's Chemical Corridor Disputes, Cambridge, MA: MIT Press.

Allen, K., Daro, V., and Holland, D.C. (2007) Becoming an Environmental Justice Activist. In R. Sandler and P.C. Pezzullo (eds.), Environmental Justice and Environmentalism: The Social Justice Challenge to the Environmental Movement, Cambridge, MA: MIT, pp. 105–135.

Allied Analytics. (2018) Waste Management Market by Waste Type and Service: Global Opportunity Analysis and Industry Forecast, 2018–2025. https://www.researchandmarkets.com/reports/4730314/waste-managementmarket-by-waste-type-and-.

Anderson, L. (2016) Inequality in children's mental health and behavioural problems: What are the trends? http://csi.nuff.ox.ac.uk/wp-content/uploads/2016/02/CSI-19-Inequalities-in-childrens-health.pdf.

Andrew, R.M. (2018) Global CO_2 emissions from cement production, Earth System Science. Data, 10: 195–217.

Antonucci, L., et al. (2017) Brexit was not the voice of the working class nor of the uneducated – it was of the squeezed middle. http://blogs.lse.ac.uk/politicsandpolicy/brexit-and-the-squeezed-middle/.

Arnstein, S.R. (1969) A Ladder of Citizen Participation. Journal of the American Planning Association 35: 216–224.

Atherton, K., and Power, C. (2007) Health inequalities with the National Statistics-Socioeconomic classification: disease risk factors and health in the 1958 British birth cohort. European Journal of Public Health, 17(5): 486–491.

Atkinson W. (2010) Class, Individualization and Late Modernity: In Search of the Reflexive Worker. Basingstoke: Palgrave.

Atkinson, W. (2015) Class. Cambridge: Polity.

Autor, D.H., and Dorn, D. (2013) The growth of low skill service jobs and the polarization of the US labour market, American Economic Review, 103(5): 1553–1597.

Baek, J., and Gweisah, G. (2013) Does income inequality harm the environment?: Empirical evidence from the United States. Energy Policy, 62: 1434–1437.

Barca, S. (2012) On Working-class Environmentalism: A Historical and Transnational Overview. Interface, 4(2): 61–80.

Barca, S., and Leonardi, E. (2016) Working-Class Communities and Ecology. In M. Shaw and M. Mayo (eds.), Class, Inequality and Community Development. Bristol: Policy Press.

Bateman, I.J., Day, B., Georgiou, S., and Lake, I. (2006) The aggregation of environmental benefit values: Welfare measures, distance decay and total WTP. Ecological Economics 60: 450–460.

Batty, E., Beatty, C., Foden, F., Lawless, P., Pearson, S., and Wilson, I. (2010) The New Deal for Communities Experience: A Final Assessment. Vol. 7. London: Department for Communities and Local Government.

Baumann, Z. (2001) The Individualized Society. Cambridge: Polity.

Baynes, C. (2019) Ocasio-Cortez delivers devastating address to congress after Republican calls Green New Deal elitist: 'People are dying', The Independent, 27 March.

BBC. (2015) Boomeco fined over Avonmouth fly infestation. https://www.bbc.co.uk/news/uk-england-bristol-33723422.

BBC. (2017) Churngold Recycling firm fined for dumping toxic waste. http://www.bbc.co.uk/news/uk-england-bristol-40441778, 6 July 2017.

BBPA. (2017) British Beer and Pub Association https://beerandpub.com/.

BCC. (2011) The Community's Vision for Lockleaze. Bristol City Council. https://www.bristol.gov.uk/documents/20182/33792/Vision%20for%20Lockleaze.pdf/3772c9e6-d318-4590-9ee1-4943c0c08a48.

BCC. (2015) Our resilient future: a framework for climate and energy security. Bristol City Council. https://www.bristol.gov.uk/documents/20182/33423/Our+Resilient+Future+A+Framework+for+Climate%20+and+Energy+Security/2ee3fe3d-efa5-425a-b271-14dca33517e6.

BCC. (2017a) Avonmouth and Lawrence Weston, Statistical Ward Profile 2017 A. https://www.bristol.gov.uk/documents/20182/436737/Avonmouth+and+Lawrence+Weston.pdf/16c104ba-fa07-43b4-b1be-4893ed77de83.

BCC. (2017b) Clifton, Statistical Ward Profile 2017. https://www.bristol.gov.uk/documents/20182/436737/Clifton.pdf/2205b281-7330-4e4c-a09a-8de549221929.

BCC. (2017c) Hartcliffe and Withywood, Statistical Ward Profile 2017. https://www.bristol.gov.uk/documents/20182/436737/Hartcliffe+and+Withywood.pdf/49d31847-00da-471c-95c8-82630662e073.

BCC. (2017d) Statistical Ward Profile Stoke Bishop. https://www.bristol.gov.uk/documents/20182/436737/Stoke+Bishop.pdf/0be3a2c1-4235-4db8-abe2-b457c8da63b2.

BCC. (2017e) Statistical Ward Profile Lawrence Hill https://www.bristol.gov.uk/documents/20182/436737/Lawrence+Hill.pdf/bec15541-2bf1-4702-9d70-c9f5d54f8bb2.

BCC. (2019) Child Poverty. https://opendata.bristol.gov.uk/explore/dataset/child-poverty-by-lsoa/information/.

Beck, U. (1992) Risk Society: Towards a New Modernity. London: Sage.

Beck, U. (1995) Ecological Politics in an Age of Risk. Cambridge: Polity.

Beck, U. (1999) World Risk Society. Oxford: Polity.

Beck, U., and Beck-Gernsheim, E. (2002) Individualization: Institutionalized individualism and its social and political consequences. London: Imprint.

Been, V. (1994) Locally Undesirable Land Uses in Minority Neighbourhoods: Disproportionate Siting or Market Dynamics? Yale Law Journal, 103(6): 1383–1422.

Beider, H. (2011) White working-class views of neighbourhood, cohesion and change. York: Joseph Rowntree Foundation.

Bell, K. (2008) Achieving Environmental Justice in the United Kingdom: A Case Study of Lockleaze. Environmental Justice 1(4): 203–210.

Bell, K. (2011a) St. Pauls Young People in 2011. Bristol: St. Pauls Unlimited/ Places for People.

Bell, K. (2011b) Environmental Justice: Lessons from Cuba. PhD. University of Bristol.

Bell, K. (2012) Is socially-just degrowth compatible with capitalism?, Degrowth Conference, Montreal 2012. https://www.degrowth.info/en/catalogue-entry/is-socially-just-degrowth-compatible-with-capitalism/.

Bell, K. (2014) Achieving Environmental Justice. Bristol: Policy Press.

Bell, K. (2015) Can the capitalist economic system deliver environmental justice? Environmental Research Letters, 10(12): 1–8.

Bell, K. (2016) Green economy or living well? Assessing divergent paradigms for equitable eco-social transition in South Korea and Bolivia. Journal of Political Ecology, 23: 71–92.

Bell, K. (2017) 'Living Well' as a path to social, ecological and economic sustainability. Urban Planning, 2(4): 19–33.

Bell, K., and Reed, M. (2019, forthcoming) The Tree of Participation: A New Tool for Engagers in Participatory Environmental Decision-Making.

Bell, K., and Sweeting, D. (2013) Urban Waste Collection – An Environmental Justice Issue? In M.J. Zapata (ed.), Waste in the City. Bristol: Policy Press: pp 201–223.

Bellamy Foster, J. (2019) John Bellamy Foster on the Green New Deal, Climate and Capitalism, 2nd Feb 2019.

Belle, D., Doucet, J., Harris, J., Miller, J., and Tan, E. (2000) Who is rich? Who is happy? American Psychologist, 55: 1160–1161.

Benn, T. (2007) If we can find the money to kill people, we can find the money to help people. Interview with Michael Moore in the film 'Sicko'.

Bennett, C. (2015) Green movement must escape its 'white, middle-class ghetto', says Friends of the Earth chief Craig Bennett. In Tim Basden, The Independent, 4th July 2015. http://www.independent.co.uk/environment/green-movement-must-escape-its-white-middle-class-ghetto-says-friends-ofthe-earth-chief-craig-10366564.html.

Bernstein, B. (1971) Class, Codes and Control. Vol. 1: Theoretical studies towards a sociology of language. London: Routledge and Kegan Paul.

Bernstein, B. (ed.). (1973) Class, Codes and Control. Vol. 2: Applied studies towards a sociology of language. London: Routledge and Kegan Paul.

Bernstein, B. (1975) Class, Codes and Control. Vol. 3: Towards a theory of educational transmissions. London: Routledge and Kegan Paul.

Bernstein, B. (1990) Class, Codes and Control. Vol. 4: The structuring of pedagogic discourse. London: Routledge and Kegan Paul.

Bihouix, P. (forthcoming) The Age of Low Tech: Towards a Technologically Sustainable Civilisation. http://www.seuil.com/ouvrage/l-age-des-lowtechphilippebihouix/9782021160727; http://www.lowtechmagazine.com/sitemap.html.

Binswanger, M. (2015) The growth imperative revisited: a rejoinder to Gilányi and Johnson. Journal of Post Keynesian Economics, 37(4): 648–660.

Bioacid. (2017) https://www.oceanacidification.de/?lang=en. BIOACID (Biological Impacts of Ocean Acidification).

Bjornsdottir, T., and Rule, N. (2017) The Visibility of Social Class From Facial Cues, Journal of Personality and Social Psychology, 113(4): 530–546.

Blyth, M. (2013) Austerity: The History of a Dangerous Idea, Oxford: OUP.

Boardman, B., et al. (1999) Equity and the Environment: Guidelines for Green and Socially Just Government, London: Catalyst Trust and Friends of the Earth (FoE).

Bolton, P. (2010) Higher education and social class. House of Commons Library: Research Briefings.

Booth, R. (2018a) Grenfell firefighters ran out of basic equipment, inquiry hears. Guardian, 25th July. https://www.theguardian.com/uk-news/2018/jul/05/grenfell-firefighters-ran-out-of-basic-equipment-inquiry-hears.

Booth, R. (2018b) Grenfell Tower: fire-resistant cladding plan was dropped. Guardian, 8th May. https://www.theguardian.com/uk-news/2018/may/08/grenfell-tower-more-costly-fire-resistant-cladding-plan-was-dropped.

Bourdieu, P. (1984) Distinction: A Social Critique of the Judgement of Taste. Cambridge, MA: Harvard University Press.

Bourdieu, P. (1990) The Logic of Practice. Redwood City, CA: Stanford University Press.

Boutelier, M., Cleverly, S., and Labonte, R. (2000) Community as a setting for health promotion. In B. Poland, L.W. Green, and I. Rootman (eds.), Settings for health promotion: linking theory and practice. Thousand Oaks, CA: Sage, pp 250–307.

Bowden, S., and Tweedale, G. (2003) Mondays without dread: The Trade Union response to byssinosis in the Lancashire cotton industry in the twentieth century, Social History of Medicine, 16(1): 79–95.

Bowling, A., Barber, J., Morris, R., and Ebrahim, S. (2006) Do perceptions of neighbourhood environment influence health? Baseline findings from a British survey of aging. Journal of Epidemiology and Community Health, 60: 476–483.

Boyle, M. (2010) The Moneyless Man: Year of Freeconomic Living. Oneworld Publications.

Boyle, A. (2012) Human Rights and the Environment: Where Next? European Journal of International Law, 23(3): 613–642.

Bramley, G., Bailey, N., Hastings, A., Watkins, D., and Croudace, R. (2012) Environmental justice in the city? Challenges for policy and resource allocation in keeping the streets clean. Environment and Planning A, 44(3): 741–761.

Braver, E.R. (2003) Race, Hispanic origin, and socioeconomic status in relation to motor vehicles occupant death rates and risk factors among adults. Accident Analysis Prevention, 35: 295–309.

Brawn, E., et al. (2013) The Other Care Crisis: Making Social Care Work for Disabled Adults in England, London.

Breen, R., and Jonsson, J. (2005) Inequality of opportunity in comparative perspective: Recent research on educational and social mobility. Annual Review of Sociology, 31: 223–243.

Bristol City Council. (2015) Deprivation in Bristol 2015. https://www.bristol.gov.uk/documents/20182/32951/Deprivation+in+Bristol+2015/429b2004-eeff-44c5-8044-9e7dcd002faf.

Bristol Green Capital Better Bristol. (2017) http://www.crowdfunder.co.uk/better-bristol-2017.

Bristol Green Capital Partnership. (2014) http://bristolgreencapital.org/.

Bristol Health and Adult Social Care Scrutiny Commission. (2015) Paper for Neighbourhood Scrutiny Commission. https://democracy.bristol.gov.uk/Data/Neighbourhoods%20Scrutiny%20Commission/201511231000/Agenda/1123_8.pdf.

Brown, A.P. (2002) Community involvement: Findings from Working for Communities. Development Department Research Programme Research Findings no. 137 Scottish Executive Research Unit.

Brown, C., Walpole, M., Simpson, L., and Tierney, M. (2011) Introduction to the UK National Ecosystem Assessment. In: The UK National Ecosystem Assessment Technical Report. UK National Ecosystem Assessment, UNEP-WCMCB.

Brulle, R. (2018) 30 years ago global warming became front-page news – and both Republicans and Democrats took it seriously, The Conversation. https://theconversation.com/30-years-ago-global-warmingbecame-front-page-news-and-both-republicans-and-democrats-took-it-seriously-97658.

Buck, D., and Maguire, D. (2015) Inequalities in life expectancy Changes over time and implications for policy, The Kings Fund. https://www.kingsfund.org.uk/sites/default/files/field/field_publication_file/inequalities-in-life-expectancy-kings-fund-aug15.pdf.

Bufton, M., and Melling, J. (2005) 'A Mere Matter of Rock': Organized Labour, Scientific Evidence and British Government Schemes for Compensation of Silicosis and Pneumoconiosis among Coalminers, 1926–1940., Medical History, 49: 155–178.

Bullard, R.D. (1983) Solid Waste Sites and the Black Houston Community. Sociological Enquiry, 53(2–3): 273–288.

Bullard, R.D. (1990) Dumping in Dixie: Race, Class, and Environmental Quality. Boulder, CO: Westview Press.

Bullard, R. (2000) Dumping in Dixie: Race, Class and Environmental Quality, Boulder, AZ: West View Press.

Bullard, R.D., and Johnson, G.S. (2000) Environmental Justice: Grassroots activism and its Impact on public policy decision making. Journal of Social Issues, 56(3): 555–578.

Bullard, R.D., and Wright, B.H. (1992) The Quest for Environmental Equity: Mobilizing the African-American Community for Social Change. In R.E. Dunlap and A.G. Mertig (eds.), American Environmentalism: The U.S. Environmental Movement, 1970–1990. New York: Taylor and Francis, pp. 39–49.

Bundred, S. (2016) Report to Bristol City Council Review of Bristol 2015 European Green Capital Year. https://www.bristol.gov.uk/documents/20182/ 1352057/European+Green+Capital+Review+report/f7ae017a-57b5-4bc0-acdf-a1ed61380a35.

Burchardt, T. (2008) Time and income poverty. Centre for Analysis of Social Exclusion, London School of Economics. http://sticerd.lse.ac.uk/dps/case/ cr/CASEreport57.pdf, downloaded 2/4/12.

Burningham, K., and Thrush, D. (2001) Rainforests are a Long Way from Here: The Environmental Concerns of Disadvantaged Groups. York: YPS for the Joseph Rowntree Foundation.

Burt, J., Stewart, D., Preston, S., and Costley, T. (2013) Monitor of Engagement with the Natural Environment Survey (2009–2012): Difference in access to the natural environment between social groups within the adult English population. Natural England Data Reports, Number 003.

Busch, F. (2015) Unemployment and insecurity in the UK labour market. Centre for Social Investigation 11: briefing note.

CAAT. (2014) Arms to Renewables: Work for the Future. https://www.caat.org. uk/campaigns/arms-to-renewables/arms-to-renewables-background-briefing.pdf.

Cai, W., Wang, C., Chen, J., and Wang, S. (2011) Green economy and green jobs: Myth or reality? The case of China's power generation sector. Energy, 36: 5994–6003.

Cameron, D. (2010) Speech to DECC, 14 May.

Cameron, D. (2012) Speech to Confederation of British Industry Annual Conference, 20 November.

Campaign Against Climate Change. (2010) One Million Climate Jobs. http:// www.campaigncc.org/greenjobs/

Campaign for Better Transport. (2018) Buses in Crisis A report on bus funding across England and Wales 2010–2018. http://bettertransport.org.uk/sites/ default/files/research-files/Buses-in-Crisis-2018_0.pdf.

Cancer Research UK. (2018) Cancer Risk Statistics. https://www.cancerresearchuk.org/health-professional/cancer-statistics/risk.

Capacity Global. (2009) Every Action Counts: The Diversity Report. London.

Carattini, S., Carvalho, M., Fankhauser, S. (2018) Overcoming public resistance to carbon taxes. WIREsClim Change 9:e531.

Carvalho, D.S., and Fidelis, T. (2009) The perception of environmental quality in Aveiro, Portugal: a study of complaints on environmental issues submitted to the City Council. Local Environment: International Journal of Justice and Sustainability, 14(10): 939–961.

Caspi, C.E., Sorensen, G., Subramanian, S.V., and Kawachi, I. (2012) The Local Food Environment and Diet: A Systematic Review. Health and Place 18: 1172–1187.

Cavlovic, T.A., Baker, K.H., Berrens, R.P., and Gawande, K. (2000) A Meta-analysis of Environmental Kuznets Curve Studies. Agricultural and Resource Economics Review 29: 32–42.

CDP. (1977) Gilding the Ghetto: The State and the Poverty Experiments. London: Community Development Projects.

Centre for Sustainable Energy. (2018) https://www.cse.org.uk/.

Cetateanu, A., and Jones A. (2014) Understanding the relationship between food environments, deprivation and childhood overweight and obesity: Evidence from a cross sectional England wide study. Health and Place 27: 68–76.

Chew, D. (1982) Ada Nield Chew: The Life and Writings of a Working Woman. London: Virago.

Chomsky, N. (1997) What Makes Mainstream Media Mainstream?, Z Magazine. www.chomsky.info/articles/199710%2D%2D.htm.

Chow, L. (2017) European Glyphosate Safety Report Copy-Pasted Monsanto Study, EcoWatch. https://www.ecowatch.com/eu-glyphosate-monsanto-2485590981.html.

Civil Society Reflection Group on the 2030 Agenda for Sustainable Development. (2018). https://www.2030spotlight.org/sites/default/files/spot2018/Spotlight_2018_web.pdf.

Clarke, L.M., Gleeson, C., and Sahin-Dikmen, M. (2019) Green Transitions in the Built Environment, University of Westminster, Working Paper 107.

CLASS. (2016) Conference, 2016 - Britain at a Crossroads: Finding the Progressive Path. Session on 'Climate Change, Green Jobs and Sustainability'.

CLASS. (2017) 'CLASS on Class' Centre for Labour and Social Studies. https://player.fm/series/class-on-class.

ClientEarth. (2017) UK facing legal action over cost rules which deter citizens from court. https://www.clientearth.org/government-facing-court-newenvironmental-justice-plans/.

Clinton, W.J. (1994) Executive Order: Federal Actions to Address Environmental Justice in Minority Populations and Low-Income Populations. Washington, DC: White House.

Cock, J., and Lambert, R. (2012) The Neo-liberal Global Economy and Nature: Redefining the Trade Union Role. In N. Räthzel and D. Uzzell (eds.), Trade Unions in the Green Economy. Working for the Environment. Oxford/New York: Earthscan/Routledge.

Cockburn, C. (2015) Standpoint Theory. In S. Mojab (ed.), Marxism and Feminism. London: Zed Books, pp. 331–346.

Cohen, R.M. (1998) Class consciousness and its consequences: The impact of an elite education on mature, working-class women. American Educational Research Journal 35: 353–375.

COMEAP. (2010) The mortality effects of long-term exposure to particulate air pollution in the United Kingdom, A report by the committee on the medical effects of air pollutants. UK: Health Protection Agency.

Committee on Fuel Povert. (2017) Membership. https://www.gov.uk/government/organisations/committee-on-fuel-poverty/about/membership.

Competition and Markets Authority. (2016) Energy market investigation Summary of final report. https://www.gov.uk/government/uploads/system/uploads/attachment_data/file/531157/Energy-final-report-summary.pdf.

Conlon, G., and Patrignani, P. (2011) The returns to higher education qualifications. BIS Research Paper Number 45.

Cook, S., Smith, K., et al. (2012) Green Economy or Green Society? Contestation and Policies for a Fair Transition, Occasional paper 10: Social Dimensions of Green Economy and Sustainable Development, United Nations Research Institute for Social Development.

Cooke, B., and Kothari, U. (eds). (2001) Participation: The New Tyranny? London: Zed Books.

Costanza, R. (2018) Wellbeing Economy, https://wellbeingeconomy.org.

Cox, S. (2018) An engineer, an economist, and an ecomodernist walk into a bar and order a free lunch, July 30th Climate and Capitalism. http://climateandcapitalism.com/2018/07/30/ecomodernist-free-lunch/.

Cozzarelli, C., Wilkinson, A.V., and Tagler, M.J. (2001) Attitudes toward the poor and attributions for poverty. Journal of Social Issues, 57(2): 207–227.

CPRE. (2018) National Parks and AONBs: access for all, says CPRE. https://www.cpre.org.uk/media-centre/latest-news-releases/item/5026-national-parks-aonbs-access-for-all-says-cpre.

Crawford, C. (2014) Socio-economic differences in university outcomes in the UK; Drop-out, degree completion and degree class, IFS working paper, W14/31.

Crawford, C., et al. (2011) Social Mobility: A Literature Review. London: Department for Business, Innovation and Skills. http://www.bis.gov.uk/assets/biscore/economics-and-statistics/docs/s/11-750-social-mobilityliterature-review.

CSDH. (2008) CSDH Final Report: Closing the gap in a generation: Health equity through action on the social determinants of health. Geneva.

CSRI. (2016) The Credit Suisse Research Institute's Global Wealth Report.

Currie, G., and Delbosc, A. (2010) Modelling the social and psychological impacts of transport disadvantage. Transportation, 37(6): 953–966.

Dalton, R.J. (2008) Citizen politics: public opinion and political parties in advanced industrial democracies, Washington, DC: CQ Press.

Dargan, L. (2004) A New Approach to Regeneration?: Reflections on the New Deal for Communities in Newcastle upon Tyne. Centre for Urban and Regional Development Studies, University of Newcastle upon Tyne. www. ncl.ac.uk/curds.

Dauvergne, P. (2016) Environmentalism of the Rich. Cambridge, MA: MIT Press.

Davis, A., Hill, K., Hirsch, D., and Padley, M. (2016) A Minimum Income Standard for the UK in 2016. York: Joseph Rowntree Foundation.

Days Group. (2017) Incinerator Bottom Ash Recycling. http://www.daygroup. co.uk/incinerator-bottom-ash-recycling.php.

Dearden, L. (2019) Rebellion: More than 1,000 protestors arrested, The Independent. April 22.

DECC. (2009) The UK Renewable Energy Strategy. London: HM Gov.

DECC. (2016) Fuel Poverty Statistics Report. https://www.gov.uk/government/ uploads/system/uploads/attachment_data/file/557400/Annual_Fuel_ Poverty_Statistics_Report_2016_-_revised_30.09.2016.pdf.

Defra. (2004) SDRN Briefing 2 – Environment and Social Justice. http://www. sd-research.org.uk/sites/default/files/publications/Environment%20 and%20Social%20Justice_0.pdf.

Defra. (2011) The Natural Choice: securing the value of nature. https://web. archive.org/web/20110725050649/http://www.official-documents.gov. uk:80/document/cm80/8082/8082.pdf.

Defra. (2013) Sustainable Development Indicators. https://www.gov.uk/gov- ernment/uploads/system/uploads/attachment_data/file/223992/0_SDIs_ final__2_.pdf.

Dempsey, N., Bramley, G., Power, S., and Brown, C. (2011) The social dimen- sion of sustainable development: defining urban social sustainability. Sustainable Development 19: 289–300.

Deutz, P. (2014) A class-based analysis of sustainable development: Developing a radical perspective on environmental justice. Sustainable Development 22: 243–252.

DfT. (2013) National Travel Survey, Department for Transport. https://www. gov.uk/government/statistics/national-travel-survey-2013.

DfT. (2017) Reported road casualties Great Britain, annual report: 2016, Department for Transport https://www.gov.uk/government/statistics/reported-road-casualties-great-britain-annual-report-2016.

Di Chiro, G. (1996) Nature as Community: The Convergence of Environment and Social Justice. In W. Cronon (ed.), Uncommon Ground: Rethinking the Human Place in Nature. New York: W.W. Norton, pp. 298–320.

Diekmann, A., and Franzen, A. (1999) The Wealth of Nations and Environmental Concern. Environment and Behavior 31(4): 540–549.

Dinham, A. (2005) Empowered or over-powered? The real experiences of local participation in the UK's New Deal for Communities. Community Development Journal, 40(3): 301–312.

Dittmar, H. (2007) The Costs of Consumer Culture and the "Cage Within": The Impact of the Material "Good Life" and "Body Perfect" Ideals on Individuals, Identity and Well-Being, Psychological Inquiry, 18(1): 23–31.

Dobson, A. (2007) Green Political Thought. London/New York: Routledge.

Dominick, R. (1998) Capitalism, communism, and environmental protection - Lessons from the German experience. Environmental History 3(3): 311–332.

Dorling, D. (2010a) Is more equal more green? London: University of Sheffield.

Dorling, D. (2010b) Social inequality and environmental justice. Environmental Scientist, 19(3): 9–13.

Dorling, D. (2015) Injustice: Why Social Inequality Still Persists. Bristol: Policy Press.

Dorling, D. (2017) Is Inequality Bad for the Environment? https://www.theguardian.com/inequality/2017/jul/04/is-inequality-bad-for-theenvironment.

Dorling, D. (2018) Peak Inequality: Britain's Ticking Time Bomb. Bristol: Policy Press.

Douthwaite, R. (1999) The Growth Illusion. How Economic Growth has Enriched the Few, Impoverished the Many and Endangered the Planet. Gabriola Island, BC: New Society Publishers.

Dover, M.A. (2016) The moment of microaggression: The experience of acts of oppression, dehumanization and exploitation. Journal of Human Behavior in the Social Environment 26(7–8): 575–586.

Drever, F., Doran, T., and Whitehead, M. (2004) Exploring the relation between class, gender, and self rated general health using the new socioeconomic classification. A study using data from the 2001 census. Journal of Epidemioly and Community Health, 58(7): 590–596.

Druckman, A., and Jackson, T. (2009) The carbon footprint of UK households 1990–2004: a socio-economically disaggregated, quasi- multiregional input-output model. Ecological Economics, 68(7): 2066–2077.

Drury, C. (2018) Mark Carney warns robots taking jobs could lead to rise of Marxism, The Independent, 14th April. https://www.independent.co.uk/news/uk/home-news/mark-carney-marxism-automation-bank-of-england-governor-job-losses-capitalism-a8304706.html.

Due, P. et al. (2009) Socioeconomic inequality in exposure to bullying during adolescence: A comparative, cross-sectional, multilevel study in 35 Countries. American Journal of Public Health, 99(5): 907–914.

Dung, T.T.T., Vassilieva, E., Golreihan, A., et al. (2017) Release of trace elements from bottom ash from hazardous waste incinerator. Recycling, 3(3): 36–54.

Dunlap, R.E., and Mertig, A.G. (1995) Global concern for the environment: Is affluence a prerequisite?. Journal of Social Issues, 51: 121–137.

Dunlap, R.E., and Mertig, A.G. (1997) Global environmental concern: An anomaly for postmaterialism. Social Science Quarterly 78(1): 24–29.

Dunlap, R.E., and York, R. (2008) The globalization of environmental concern and the limits of the postmaterialist values explanation: Evidence from four multinational surveys. Sociological Quarterly, 49: 529–563.

Easterlin, R.A. (2013) Happiness, growth and public policy. Economic Inquiry, 51(1): 1–15.

Easterlin, R.A., McVey, L.A., Switek, M., Sawangfa, O., and Zweig, J.S. (2010) The happiness-income paradox revisited. Proceedings of the National Academy of Sciences, 107(52): 22463–22468.

Eckenrode, J., et al. (2014) Income inequality and child maltreatment in the United States. Pediatrics, 133(3): 454–461.

Edwards, P., et al. (2007) Deaths from injury in children and employment status in family: analysis of trends in class specific death rates, BMJ, 333: 119–121.

Egan, M., Katikireddi, S.V., Kearns, A., Tannahill, C., Kalacs, M., and Bond, L. (2013) Health effects of neighborhood demolition and housing improvement: a prospective controlled study of 2 natural experiments in urban renewal. American Journal of Public Health, 103(6): e47–e53. pmid: 23597345.

EHRC. (2016) Race report: Healing a divided Britain. Equality and Human Rights Commission. https://www.equalityhumanrights.com/en/race-reporthealing-divided-britain.

Ekins, P., and Lockwood, M. (2011) Tackling Fuel Poverty During the Transition to a Low-Carbon Economy. York: Joseph Rowntree Foundation.

Ellaway, A., Morris, G., Curtice, J., Robertson, C., Allardice, G., and Robertson, R. (2009) Associations between health and different types of environmental incivility: a Scotland-wide study. Public Health, 123: 708–713.

Ellaway, A., Benzeval, M., Green, M., Leyland, A., and Macintyre, S. (2012) Getting sicker quicker: Does living in more deprived neighbourhood mean your health deteriorates faster? Health and Place, 18(2): 132–137.

Elliot, R., and Leonard, C. (2006) Peer pressure and poverty: exploring fashion brands and consumption symbolism among children of the 'British poor'. Journal of Consumer Behaviour, 3(4): 347–359.

End Child Poverty. (2016) http://www.endchildpoverty.org.uk/povertyin-your-area-2016/.

Engels, F. (1845/1967) The Condition of the Working-class in England. Panther Press.

Environment Agency. (2007) Using Science to Create a Better Place: Addressing environmental inequalities: cumulative environmental impacts. Bristol: Environment Agency.

Environment Agency. (2015a) Earthminded (Avonmouth) 3rd Feb 2015 Blow Pink Smoke all over Avonmouth. https://issuu.com/bristolnews/docs/car_form_2015_02_03_incident.

Environment Agency. (2015b) Avonmouth: Dust Monitoring. https://www.gov.uk/government/publications/avonmouth-fly-and-dust-issues/avonmouth-dust-monitoring.

Environment Agency. (2017a) What's in Your Backyard. http://apps.environ-ment-agency.gov.uk/wiyby/default.aspx.

Environment Agency. (2017b) Environment Agency permitting decisions. https://assets.publishing.service.gov.uk/government/uploads/system/uploads/attachment_data/file/608610/Avonmouth_IBA_Refusal_DD.pdf.

Environment Agency. (2017c) Refusal of an application for a permit. Application no. EPR/TP3138DP/A001. https://assets.publishing.service.gov.uk/govern-ment/uploads/system/uploads/attachment_data/file/608608/Avonmouth_IBA_Refusal_notice.pdf.

Environment Agency. (2018) Day Group Limited: environmental permit con-sultation. https://consult.environment-agency.gov.uk/psc/day-group-avon-mouth/consultation/published_select_respondent.

Equality Act. (2010) http://www.legislation.gov.uk/ukpga/2010/15/contents.

Erikson, R., Goldthorpe, J.H., Jackson, M., Yaish, M., and Cox, D.R. (2005) On class differentials in educational attainment. Proceedings of the National Academy of Sciences 102(27): 9730–9733.

Erlich, P. (2018) in Carrington, D.: Paul Ehrlich: 'Collapse of civilisation is a near certainty within decades'. The Guardian, 22nd March. https://www.the-guardian.com/cities/2018/mar/22/collapse-civilisation-near-certaindecades-population-bomb-paul-ehrlich.

Evans, G., and Tilley, J. (2017) The New Politics of Class. Oxford: OUP.

Faber, D. (2008) Capitalizing on Environmental Injustice: The Polluter Industrial Complex in the Age of Globalization. Lanham, MD: Rowman and Littlefield.

Factor, R., Yair, G., and Mahalel, D. (2010) Who by accident? The social morphology of car accidents. Risk Analysis, 30: 1411–1423.

Fairburn, J., Butler, B., and Smith, G. (2009) Environmental justice in South Yorkshire: Locating social deprivation and poor environments using multiple indicators. Local Environment, 14(2): 139–154.

FAO. (2015) The state of food insecurity in the world 2015. Food and Agriculture Organisation of the United Nations. www.fao.org/hunger/en/.

Farber, H.S., Herbst, D., Kuziemko, I., and Naidu, S. (2018) Unions and Inequality Over the Twentieth Century: New Evidence from Survey Data, NBER Working Paper No. 24587, NBER Program(s): Development of the American Economy, Labor Studies, Political Economy, The National Bureau of Economic Research. http://www.nber.org/papers/w24587.

Feeney, B.C., and Collins, N.L. (2015) A New Look at Social Support: A Theoretical Perspective on Thriving Through Relationships. Personality and Social Psychology Review, 19(2): 113–147.

Fell, D. (2016) Bad Habits, Hard Choices: Using the Tax System to Make Us Healthier. London: London Publishing Partnership.

Feola, G., and Nunes, R.J. (2013) Failure and Success of Transition Initiatives: A Study of the International Replication of the Transition Movement (Research Note 4). Reading: Walker Institute for Climate System Research.

FGBP. (2017) Filwood Green Business Park. https://filwoodgreen.co.uk/about.

FIA. (2017) https://www.fiafoundation.org/.

Fielding, J., and Burningham, K. (2005) Environmental inequality and flood Hazard. Local Environment, 10(4): 1–17.

First National People of Color. (1991) Principles of Environmental Justice First Nation People of Color Environmental Leadership Summit, Washington, DC.

Fisher, S. (2013) Our sustainability crisis didn't start and doesn't stop at climate change. The Conversation.

Fiske, S.T., and Markus, H.R. (2012) Facing Social Class: How Societal Rank Influences Interaction. New York: Russell Sage.

Fiske, S.T., Moya, M., Russell, A.M., and Bearns, C. (2012) The secret handshake: Trust in cross-class encounters. In S.T. Fiske and H.R. Markus (eds.), Facing Social Class: How societal rank influences interaction. New York: Russell Sage, pp 234–252.

Fitzpatrick, T. (ed.). (2011) Understanding the Environment and Social Policy. Bristol: Policy Press.

Fitzpatrick, T. (ed.). (2014a) International Handbook on Social Policy and the Environment. Cheltenham: Edward Elgar.

Fitzpatrick, T. (2014b) Climate Change and Poverty: A New Agenda for Developed Nations. Bristol: Policy Press.

FoE. (2000) Pollution Injustice. London: Friends of the Earth.

FoE. (2001) Pollution and Poverty – Breaking the Link. London: Friends of the Earth.

FoE. (2002) The safety of incinerator ash, Friends of the Earth. https://friendsoftheearth.uk/sites/default/files/downloads/safety_incinerator_ash.pdf.

FoE. (2004) Incinerators and Deprivation. www.foe.co.uk/resource/briefings/incineration_deprivation.pdf.

FoEs. (2017) An Environmental Court or Tribunal for Scotland, Friends of the Earth, Scotland.

Foley, L., Prins, R., Crawford, F., Humphreys, D., Mitchell, R., Sahlqvist, S., et al. (2017) Effects of living near an urban motorway on the wellbeing of local residents in deprived areas: Natural experimental study. PLoS ONE 12(4): e0174882.

Font, A., and Fuller, G. (2017) A Return to Winter Smog? Kings College London Air Quality Network (LAQN) Conference, 13th July 2017.

Forbes. (2017) The World's Billionaires. https://www.forbes.com/billionaires/list/.

Foster, J.B. (2008) Ecology and the Transition from Capitalism to Socialism. New York: Monthly Review Press.

Foster, J.B. (2013) The Epochal Crisis, 5(65). https://monthlyreview.org/2013/10/01/epochal-crisis/.

Franzen, A. (2003) Environmental Attitudes in International Comparison: An Analysis of the ISSP Surveys 1993 and 2000. Social Science Quarterly, 84(2): 297–308.

Franzen, A., and Meyer, R. (2010) Environmental Attitudes in Cross-National Perspective: A M.

Frey, C.B., et al. (2017) Technology at Work v3.0 Automating e-Commerce from Click to Pick to Door. Citi GPS: Global Perspectives & Solutions August. https://ir.citi.com/VRwNwIr6xCMPr8cKW9DJ84mVqV0oogApl EDtFSDnMqrArP34dqqVCd1DgSV%2FP%2BFasRK5yg5Ghpc%3D.

Fritsche, I., Moya, M., Bukowski, M., Jugert, P., de Lemus, S., Decker, O., Valor-Segura, I., and Navarro-Carrillo, G. (2017) The great recession and group-based control: Converting personal helplessness into social class ingroup trust and collective action. Journal of Social Issues, 73(1): 117–137.

Fuchs, C. (2017) Critical Social Theory and Sustainable Development: The Role of Class, Capitalism and Domination in a Dialectical Analysis of Un/ Sustainability. Sustainable Development, 25(5): 443–458.

Gal, J. (1998) Formulating the Matthew Principle: on the role of the middle-classes in the welfare state. Scandinavian Journal of Social Welfare, 7: 42–55.

Gallan, P. (2018) In Dodd, V. Rising crime is symptom of inequality, says senior Met chief, Guardian June 4. https://www.theguardian.com/uknews/2018/ jun/14/rising-is-symptom-of-inequality-says-senior-metchief?platform=hoot suite.

Gallie, D. (2015) Class Inequality at Work: Trends to Polarization? In A. Felstead, D. Gallie, and F. Green (eds.), Unequal Britain at Work. Oxford: Oxford University Press, pp 22–41.

Gambhir, A., et al. (2018) Towards a just and equitable low-carbon energy transition, Grantham Institute, Briefing Paper no. 26.

Garthwaite, K. (2016) Hunger Pains: Life Inside Foodbank Britain. Bristol: Policy Press.

Gentleman, A. (2014) Vulnerable man starved to death after benefits were cut, The Guardian, 28th February 2014.

Gibson-Graham, J.K. (2006) Postcapitalist Politics. Minneapolis: University of Minnesota Press.

Giddens, A. (1991) Modernity and Self-Identity. Cambridge: Polity.

Gilbert, N. (ed.). (2001) Researching Social Life. London: Sage.

Giljum, S., Dittrich, M., Lieber, M., and Lutter, S. (2014) Global Patterns of Material Flows and their Socio-Economic and Environmental Implications: A MFA Study on All Countries World-Wide from 1980 to 2009.Resources, 2(1): 319–339.

Gingrich, M. (2012) From Blue to Green: A Comparative Study Of Blue-Collar Unions' Reactions To The Climate Change Threat in the United States and Sweden. In N. Räthzel and D. Uzzell (eds.), Trade Unions in the Green Economy. Working for the Environment. Oxford/New York: Earthscan/ Routledge.

Glass, R. (1964) Introduction: aspects of change. In London: Aspects of Change, ed. Centre for Urban Studies, London: MacKibbon and Kee, xiii–xlii.

Goeminne, P.C., Cox, B., Finch, S., et al. (2018) The impact of acute air pollution fluctuations on bronchiectasis pulmonary exacerbation: a case-crossover analysis. European Respiratory Journal, 52(1): 1702557.

Goldthorpe, J.H., and McKnight, A. (2006) The economic basis of social class. In S.L. Morgan, D.B. Grusky, and G.S. Fields (eds.), Mobility and Inequality: Frontiers of Research in Sociology and Economics. Stanford: Stanford University Press, pp. 109–136.

Goldthorpe, J.H., and Chan, Tak Wing. (2007) Class and Status: The Conceptual Distinction and its Empirical Relevance. American Sociological Review, 72(4): 512–532.

Goodall, H., and Cadzow, A. (2010) The People's National Park: Working-Class Environmental Campaigns on Sydney's Georges River, 1950–67. Labour History, 17–35.

Goodman, D.J. (2011) Promoting Diversity and Social Justice: Educating People from Privileged Groups. London: Sage.

Gore, A. (2000) Earth in the Balance: Ecology and the Human Spirit, Boston, MA: Houghton Mifflin.

Gorelick, S. (2017) Our obsolescent economy, Local Futures. https://www.localfutures.org/our-obsolescent-economy/#_edn5.

Gough, I. (2013) Carbon Mitigation Policies, Distributional Dilemmas and Social Policies. Journal of Social Policy, 42: 191–213.

Gough, I. (2017) Heat, Greed and Human Need. Climate Change, Capitalism and Sustainable Wellbeing. Cheltenham: Edward Elgar.

Goulson, D. (2017) in Carrington, D. Warning of 'ecological Armageddon' after dramatic plunge in insect numbers. Guardian 18th October 2017.

Gramsci, A. (2010) Prison Notebooks. New York: Columbia University Press.

Grandjean, P. (2013) Only One Chance: How Environmental Pollution Impairs Brain Development – and How to Protect the Brains of the Next Generation. New York: Oxford University Press.

Grant, W.B. (2009) Air Pollution in Relation to U.S. Cancer Mortality Rates: An Ecological Study; Likely Role of Carbonaceous Aerosols and Polycyclic Aromatic Hydrocarbons. Anticancer Research 29(9): 3537–3545.

Green Party. (2011) Autumn Conference. How to increase working-class representation in the Green Party. https://www.greenparty.org.uk/assets/files/conference/2011/Autumn.

Green Party. (2013) Green Vote Against Budget cuts. https://bristolgreenparty.org.uk/greens-vote-against-budget.

Green Party. (2018) A Burning Problem; How Incineration is stopping recycling. https://www.scribd.com/document/383927762/Green-party-reporton-incineration-and-recycling.

Greenpeace. (2017) China's plastic scrap ban threatens 'crisis' for UK recycling industry. https://unearthed.greenpeace.org/2017/12/07/china-plastic-scrapban-crisis-uk-recycling/.

Grenfell Tower Inquiry. (2018) https://www.grenfelltowerinquiry.org.uk/evidence.

Griffin, P. (2017) The Carbon Majors Database: CDP Carbon Majors Report 2017, CDP: London. https://b8f65cb373b1b7b15feb-c70d8ead6ced-550b4d987d7c03fcdd1d.ssl.cf3.rackcdn.com/cms/reports/documents/000/002/327/original/Carbon-Majors-Report-2017.pdf?1499691240.

Grossmann, M., and Creamer, E. (2017) Assessing diversity and inclusivity within the Transition movement: an urban case study, Environmental Politics, 26(1): 161–182.

Guardian. (2013) Little solidarity over the question of social class, April 5th. https://www.theguardian.com/society/2013/apr/05/solidarity-questionsocial-class.

Guerrero, A.M., Bennett, N.J., Wilson, K.A., et al. (2018) Achieving the promise of integration in social-ecological research: a review and prospectus. Ecology and Society, 23(3): 38.

Hadler, M. (2005) Why do people accept different income ratios? A multi-level comparison of thirty countries. Acta Sociologica, 48(2): 131–154.

Hagger, M.S., Wood, C., Stiff, C., and Chatzisarantis, N.L. (2010) Ego depletion and the strength model of self-control: a meta-analysis. Psycholical Bulletin, 136(4): 495–525.

Hall, S., Leary, K., and Greevy, H. (2014) Public Attitudes to Poverty. York: JFK.

Hallmann, C.A., Sorg, M., Jongejans, E., Siepel, H., Hofland, N., Schwan, H., et al. (2017) More than 75 percent decline over 27 years in total flying insect biomass in protected areas. PLoS ONE, 12(10): e0185809.

Halpern, D., Bates, C., and Beales, G. (2003) Personal Responsibility and Behaviour Change, Strategic Audit Paper, London: Cabinet Office Strategy Unit.

Hampton, P. (2015a) Trade unions and climate politics: prisoners of neoliberalism or swords of climate justice? 6 March. Paper presented to the Political Studies Association Conference 2015, Sheffield, 30 March 2015.

Hampton, P. (2015b) Workers and Trade Unions for Climate Solidarity: Tackling climate change in a neoliberal world. London: Routledge.

Hanley, L. (2016) Respectable: The Experience of Class. London: Penguin.

Harris, T., and Fiske, S.T. (2006) Social groups that elicit disgust are differentially processed in mPFC, Social Cognitive and Affective Neuroscience, 2(1): 45–51.

Hart, J. (2008) Driven to excess—impacts of motor vehicle traffic on residential quality of life in Bristol, UK. Bristol: University of the West of England.

Harter, J.-H. (2004) Environmental Justice for Whom? Class, New Social Movements and the Environment: A Case Study of Greenpeace Canada, 1971–2000. Labour/Le Travail, 54: 83–116.

Harvey, D. (2006) Spaces of Global Capitalism. London/New York: Verso.

Hasselberg, M., Vaez, M., and Laflamme, L. (2005) Socioeconomic aspects of the circumstances and consequences of car crashes among young adults. Social Science and Medicine, 60: 287–295.

Hastings, A. (2007) Territorial justice and neighbourhood environmental services: a comparison of provision to deprived and better off neighbourhoods in the UK. Environment and Planning C-Government and Policy, 25(6): 896–917.

Hastings, A. (2009) Poor neighbourhoods and poor services: evidence on the 'rationing' of environmental service provision to deprived neighbourhoods. Urban Studies, 46(13): 2907–2927.

Hastings, A., Flint, J., McKenzie, C., and Mills, C. (2005) Cleaning up Neighbourhoods: Environmental Problems and Service Provision in Deprived Areas. Bristol: The Policy Press.

Hastings, A., Bailey, N., Bramley, G., Croudace, R., and Watkins, D. (2014) 'Managing' the Middle-classes: Urban Managers, Public Services and the Response to Middle-Class Capture. Local Government Studies, 40(2): 203–223.

Hattersley, R. (2007) Kinder Mass Tresspass. http://www.kindertrespass.com/.

Hatzenbuehler, M.L., Phelan, J.C., and Link, B.G. (2013) Stigma as a fundamental cause of population health inequalities. American Journal of Public Health, 103(5): 813–821.

Hawkins, T.R., et al. (2013) Comparative environmental life cycle assessment of conventional and electric vehicles. Journal of Industrial Ecology, 17: 53–64.

Hazards 97, 99. http://www.hazards.org/haz109/h109unioneffect.pdf.

Helm, D. (2012) The Carbon Crunch: How We're Getting Climate Change Wrong – and How to Fix it. London: Yale University Press.

Helm, D. (2017) Burn Out – The endgame for fossil fuels. London: Yale University Press.

Henriksson, L. (2011) Slutkört. Stockholm: Ordfront.

Hey, D. (2011) Kinder Scout and the legend of the Mass Trespass. Agricultural History Review, 59(2): 199–216.

Hiam, L., Dorling, D., Harrison, D., and McKee, M. (2017) Why has mortality in England and Wales been increasing? An iterative demographic analysis. Journal of the Royal Society of Medicine, 110(4): 153–162.

Hickel, J. (2015) Five reasons to think twice about the UN's sustainable development goals, Africa at LSE, 23rd Sept 2015, blog.

Hickel, J. (2018) Why Growth Can't be Green. https://www.jasonhickel.org/blog/2018/9/14/why-growth-cant-be-green.

Hirsch, F. (1977 [2010]) The Social Limits to Growth. In Hughes, B. The social barriers to sustainability. http://www.dustormagic.net/EqualityWhen/SocialBarriersToSustainability.html.

Hoffman, M. (2017) Change put to work: A degrowth perspective on unsustainable work, postwork alternatives and politics, Master Thesis Series in Environmental Studies and Sustainability Science. http://lup.lub.lu.se/luur/download?func=downloadFile&recordOId=8904049&fileOId=8904056.

Hopkins, R. (2011) The Transition Companion. Totnes: Green Books.

Horwitz, S., Shutts, K., and Olson, K.R. (2014) Social Class Differences Produce Social Group Preferences. Developmental Science 17(6): 991–1002.

Howie, R. (2012) Asbestos – Consequences of Inspectorate Failures. Presentation at Occupational Hygiene Conference 2012, Cardiff, BOHS.

Hughes, A., and Kumari, M. (2017) Unemployment, underweight, and obesity: Findings from Understanding Society (UKHLS). Preventive Medicine, 97: 19–25.

IARC. (2015) International Agency for Research on Cancer Monographs, Volume 112: Evaluation of five organophosphate insecticides and herbicides. https://www.iarc.fr/en/media-centre/iarcnews/pdf/MonographVolume112.pdf.

ILO. (2018) International Labour Organisation, World Employment and Social Outlook 2018: Greening with Jobs, ILO. http://www.ilo.org/global/aboutthe-ilo/newsroom/news/WCMS_628644/lang–en/index.htm.

ILO and UNEP. (2012) Working towards sustainable development: Opportunities for decent work and social inclusion in a green economy. Geneva: ILO.

IMF. (2015) Dabla-Norris, E., Kochhar, K., Suphaphiphat, N., Ricka, F., and Tsounta, E., Causes and consequences of income inequality: A global perspective, International Monetary Fund. http://www.imf.org/external/pubs/ft/sdn/2015/sdn1513.pdf.

Inglehart, R. (1977) The Silent Revolution: Changing Values and Political Styles Among Western Publics. Princeton, NJ: Princeton University Press.

Inglehart, R. (1990) Culture Shift in Advanced Industrial Society. Princeton, NJ: Princeton University Press.

Inglehart, R. (1995) Public Support for Environmental Protection: Objective Problems and Subjective Values in 43 Societies. PS: Political Science and Politics, 28: 57–72.

Inglehart, R. (1997) Modernization and Postmodernization: Cultural, Economic, and Political Change in 43 Societies. Princeton, NJ: Princeton University Press.

Inglehart, R. (2000) Globalization and Postmodern Values. Washington Quarterly, 23(1): 215–228.

Institute for Global Prosperity. (2017) Social prosperity for the future: A proposal for Universal Basic Services. https://www.ucl.ac.uk/bartlett/igp/sites/bartlett/files/universal_basic_services_-_the_institute_for_global_prosperity_.pdf.

Intergovernmental Panel on Climate Change. (2018) Global Warming of 1.5 °C. http://www.ipcc.ch/report/sr15/.

IPBES. (2018) Regional Assessment Reports, Intergovernmental Science-Policy Platform on Biodiversity and Ecosystem Services, Bonn. https://www.dropbox.com/sh/9cu6sv14gik0k9m/AADIAYeijq2MlSvI-9GGelR_a?dl=0.

IPPR. (2017) Wealth in the twenty-first century: Inequalities and drivers, Report Economy Commission on Economic Justice IPPR. https://ippr.org/research/publications/wealth-inthe-twenty-first-century.

IPPR. (2018a) Institute for Public Policy Research, Beyond Eco – the Future of Fuel Poverty Support. https://www.ippr.org/files/2018-07/fuel-poverty-june18-summary.pdf.

IPPR. (2018b) Power to the People: How stronger unions can deliver economic justice. https://www.ippr.org/files/2018-06/cej-trade-unions-may18-.pdf.

Isenberg, N. (2016) The 400-Year Untold History of Class in America. New York: Penguin.

Ising, H., and Kruppa, B. (2004) Health effects caused by noise: evidence in the literature from the past 25 years. Noise and Health 6(22): 5–13.

ITUC. (2015) Call for Dialogue: Climate action requires just transition. https://www.ituc-csi.org/call-for-dialogue-climate-action.

Jackson, T. (2005) Motivating Sustainable Consumption a review of evidence on consumer behaviour and behavioural change a report to the Sustainable Development Research Network, January 2005. http://www.sustainablelifestyles.ac.uk/sites/default/files/motivating_sc_final.pdf.

Jackson, T. (2006) Beyond The 'Wellbeing Paradox': - wellbeing, consumption growth and sustainability. Centre for Environmental Strategy, University of Surrey.

Jackson, M. (2009a) Disadvantaged through Discrimination? The Role of Employers in Social Stratification. British Journal of Sociology, 60(4): 669–692.

Jackson, T. (2009b) Prosperity without Growth: Economics for a Finite Planet. London: Earthscan.

Jackson, S. (2011) 'SIPRI assessment of UK arms export subsidies', report for Campaign Against Arms Trade, 25 May 2011. bit.ly/1ttzGja.

Jackson, T. (2017) The future of jobs: is decent work for all a pipe dream 15th August 2017. https://www.theguardian.com/global-development-professionals-network/2017/aug/15/the-future-of-jobs-is-decent-workfor-all-a-pipe-dream.

Jakobsson, N., Muttarak, R., and Schoyen, M.-A. (2017) Dividing the pie in the eco-social state: Exploring the relationship between public support for environmental and welfare policies. Environment and Planning C: Politics and Space, 36(2): 313–339.

James, O. (2007) Evaluating the expectations disconfirmation and expectations anchoring approaches to citizen satisfaction with local public services. Journal of Public Administration Research and Theory, 19: 107–123.

Järvensivu, P., Toivanen, T., Vadén, T., Lähde, V., Majava, A., and Eronen, J. (2018) Invited background document on economic transformation, to chapter: Transformation: The Economy, UN Global Sustainable Development Report 2019. https://bios.fi/bios-governance_of_economic_transition.pdf.

Jenkins, K. (2018) Setting energy justice apart from the crowd: Lessons from environmental and climate justice. Energy Research and Social Science, 39: 117–121.

Jensen, B. (2012) Reading Classes: On Culture and Classism in America. Ithaca, NY: Cornell University Press.

Jevons, W.S. (1866) The Coal Question: An Inquiry Concerning the Progress of the Nation, and the Probable Exhaustion of Our Coal-Mines, London: Macmillan and Co.

Johnson, B., and Al-Hamad, A. (2011) Trends in socio-economic inequalities in female mortality, 2001–08. Intercensal estimates for England and Wales. Health Statistics Quarterly 52(1): 3–32.

Johnson, C., Penning-Rowsell, E., and Parker, D. (2007) Natural and imposed injustices: the challenges in implementing "fair" flood risk management policy in England. The Geographical Journal, 173(4): 374–390.

Jones, O. (2011) Chavs: The Demonization of the Working Class. London: Verso Books.

Joseph Rowntree Foundation. (2018) A Minimum Income Standard for the UK 2008–2018: continuity and change by Abigail Davis, Donald Hirsch, Matt Padley and Claire Shepherd. York: JRF.

Jun, Y., Zhong-kui, Y., and Peng-fei, P. (2011) Income distribution, human capital and environmental quality: empirical study in China Energy Procedia, 5: 1689–1696.

Kaplan, S. (2000) Human Nature and Environmentally Responsible Behavior. Journal of Social Issues, 56(3): 491–508.

Kawachi, I., Kennedy, B.P., Lochner, K., and Prothrow-Stith, D. (1997) Social capital, income inequality, and mortality. American Journal of Public Health, 87(9): 1491–1498.

Kenyon, S. (2006) The 'accessibility diary': Discussing a new methodological approach to understand the impact of Internet use upon personal travel and activity participation. Journal of Transport Geography, 14(2): 123–134.

Kettunen, J., Lanki, T., Tiittanen, P., Aalto, P.P., Koskentalo, T., Kulmala, M., Salomaa, V., and Pekkanen, J. (2007) Associations of Fine and Ultrafine Particulate Air Pollution With Stroke Mortality in an Area of Low Air Pollution Levels. Stroke 38: 918–922.

Kheifets, L., et al. (2010) A Pooled Analysis of Extremely Low-Frequency Magnetic Fields and Childhood Brain Tumors. American Journal of Epidemiology, 172(7): 752–761.

Kim, C., S.H. Jung, et al. (2010) Ambient Particulate Matter as a Risk Factor for Suicide. American Journal of Psychiatry, 167(9): 1100–7.

Kincheloe, J.L., and McLaren, P. (2000) Rethinking critical theory and qualitative research. In N. Denzin and Y. Lincoln (eds.), Handbook of Qualitative Research. Thousand Oaks, CA: Sage Publications, Inc., pp. 279–314.

Kiniyalocts, M. (2000) Environmental Justice: Avoiding the Difficulty of Proving Discriminatory Intent in Hazardous Waste Siting Decisions, North America Series, L.T. Center, University of Wisconsin-Madison.

Kirby, E. (2011) "But society is beyond ___ism" (?): Teaching how differences are "organized" via institutional privilege↔oppression. In Reframing Difference in Organizational Communication Studies: Research, Pedagogy, Practice. Thousand Oaks, CA: SAGE Publications Inc., pp. 127–150.

Kirby, S., and Riley, R. (2008) The external returns to education: UK evidence using repeated cross-sections. Labour Economics, 15(4): 619–630.

Klein, N. (2014) This Changes Everything: Capitalism vs. the Climate. New York: Simon & Schuster.

Kleiner, D. (2016) Universal Basic Income is a neoliberal plot to make you poorer. https://www.opendemocracy.net/neweconomics/universal-basicincome-is-a-neoliberal-plot-to-make-you-poorer/.

Knight, K.W., and Messer, B.L. (2012) Environmental Concern in Cross-National Perspective: The Effects of Affluence, Environmental Degradation, and World Society. Social Science Quarterly 93: 521–537.

Knox, G. (2008) Atmospheric Pollutants and Mortalities in English Local Authority Areas. Journal of Epidemiological Community Health, 62: 442–447.

Kovel, J. (2002) The Enemy of Nature: The End of Capitalism or the End of the World. London: Zed Books.

Krauss, C. (1993) Women And Toxic Waste Protests: Race, Class And Gender As Resources Of Resistance. Qualitative Sociology, 16(3): 247–262.

Krieger, N. (2014) Discrimination and health inequities. International Journal of Health Services, 44(4): 643–710.

Kruize, H., Droomers, M., van Kamp, I., and Ruijsbroek, A. (2014) What Causes Environmental Inequalities and Related Health Effects? An Analysis of Evolving Concepts. International Journal of Environmental Research and Public Health, 11(6): 5807–5827.

Labor Network for Sustainability. (2018) Just Transition – What it is. http://www.labor4sustainability.org/uncategorized/just-transition-just-what-is-it/.

Lakey, G. (2010) Facilitating Group learning: Strategies for Success with Diverse Adult Learners. San Francisco, CA: Jossey-Bass.

Lamont, M. (2000) The Dignity of Working Men: Morality and the Boundaries of Race, Class and Immigration. Cambridge MA: Harvard University Press.

Landrigan, Philip J., et al. (2017) The Lancet Commission on pollution and health. http://www.thelancet.com/journals/lancet/article/PIIS0140-6736(17)32345-0/fulltext.

Langford, A., and Johnson, B. (2010) Trends in social inequalities in male mortality, 2001–08. Intercensal estimates for England and Wales. Health Statistics Quarterly, 47: 5–32.

Latouche, S. (2009) Farewell to Growth. Cambridge, UK: Polity Press.

Laville, S. (2019) Ella Kissi-Debrah: new inquest granted into 'air pollution' death, 2nd May 2019.

Lawn, P., and Clarke, M. (2010) The end of economic growth? A contracting threshold hypothesis. Ecological Economics 69: 2213–2223.

Lawrence, M.J., et al. (2015) The effects of modern war and military activities on biodiversity and the environment, Environmental Review, 23(4): 443–446.

Laxen, D., Kirk-Lloyd, F., and Beattie, C. (2017) Health Impacts of Air Pollution in Bristol. Bristol: Air Quality Consultants.

Le Grand, J. (1982) The Strategy of Equality: Redistribution and the Social Services. London: George Allen and Unwin.

Legatum Institute and Populus. (2017) https://www.theguardian.com/business/2017/oct/01/jeremy-corbyn-nationalisation-plans-voters-tired-freemarkets.

Lerner, S. (2005) Diamond: A Struggle for Environmental Justice in Louisiana's Chemical Corridor. Cambridge, MA: MIT Press.

Levin, S., and Gillam, C. (2018) 'The world is against them': new era of cancer lawsuits threaten Monsanto. The Guardian, 18th October 2018. https://www.theguardian.com/science/2018/oct/07/monsanto-trialcancer-appeal-glyphosate-chemical.

Levine, P. with Frederick, A. (1997) Waking the Tiger: Healing Trauma. Berkeley, CA: North Atlantic Books.

Lewis, A., Turton, P., and Sweetman, T. (2009) Litterbugs: How to Deal with the Problem of Littering. London: Policy Exchange.

Liu, Y., and Keleher, T. (2009) The Green Equity Toolkit: Standards and Strategies for Advancing Race, Gender and Economic Equity in the Green Economy, Applied Research Center. https://www.raceforward.org/system/files/pdf/reports/Green_Toolkit_112009.pdf.

Long, V. (2011) The Rise and Fall of the Healthy Factory. Basingstoke: Palgrave.

Loomis, E. (2016) Towards a Working-Class Environmentalism, News Republic. https://newrepublic.com/article/139132/towards-working-classenvironmentalism.

Lott, B. (2002) Cognitive and Behavioral Distancing from the Poor. American Psychologist, 57(2): 100–110.

Lowe, C., Whitfield, G., Sutton, L., et al. (2011) Road user safety and disadvantage. Department for Transport. Road Safety Research Report No. 123.

Lowy Institute. (2018) The future of work. https://www.lowyinstitute.org/theinterpreter/future-work.

Lucas, K. (2012) Transport and social exclusion: Where are we now? Transport Policy, 20: 105–113.

Lucas, K., Walker, G., Eames, M., Fay, H., and Poustie, M. (2004) Environment and Social Justice: Rapid Research and Evidence Review. SDRN, Policy Studies Institute.

Maas, J., Verheij, R.A., Groenewegen, P.P., et al. (2006) Green space, urbanity, and health: how strong is the relation? Journal of Epidemiology and Community Health, 60: 587–592.

Maas, J., Verheij, R.A., de Vries, S., et al. (2009) Morbidity is related to a green living environment. Journal of Epidemiology and Community Health 63(12): 967–973.

Mac Sheoin, T. (2012) Power imbalances and claiming credit in coalition campaigns: Greenpeace and Bhopal, Interface: a journal for and about social Movements, 4(2): 490–511.

Macintyre, S., Macdonald, L., and Ellaway, A. (2008) Do poorer people have poorer access to local resources and facilities The distribution of local resources by area deprivation in Glasgow, Scotland. Social Science and Medicine, 67: 900–914.

MacPhee, K. (2014) Canadian Working-Class Environmentalism, 1965–1985. Labour-Le Travail, 123-+.

Magdoff, F., and Foster, J.B. (2011) What Every Environmentalist Needs to Know About Capitalism. New York: Monthly Review Press.

Manser, R. (1993) Failed Transitions: The Eastern European Economy and Environment Since the Fall of Communism. New York: The New Press.

Marable, M. (1983) How Capitalism Underdeveloped Black America. Boulder, CO: Westview Press.

Marmot, M. (2010) Fair Society, Healthy Lives. Report to the Department of Health. http://www.marmotreview.org/13.

Marmot, M. (2017) http://www.instituteofhealthequity.org/resources-reports/ marmot-indicators-2017-institute-of-health-equity-briefing/marmot-indica- tors-briefing-2017-updated.pdf.

Marmot, M., et al. (2010) Fair Society, Healthy Lives: A Strategic Review of Health Inequalities in England Post-2010, UCL, London.

Martinez-Alier, J. (1995) The environment as a luxury good or "too poor to be green"? Ecological Economics 13: 1–10.

Martinez-Alier, J. (2002a) The Environmentalism of the Poor: A Study of Ecological Conflicts and Valuation. Cheltenham, UK: Edward Elgar.

Martinez-Alier, J. (2002b) The Environmentalism of the Poor. University of Witswatersrand, Report for UNRISD.

Martinez-Alier, J. (2003) Environmentalism of the Poor, Basingstoke: Edward Elgar.

Martuzzi, M., Mitis, F., and Forastiere, F. (2010) Inequalities, inequities, envi- ronmental justice in waste management and health. European Journal of Public Health, 20(1): 21–26.

Marx, K. (1852/1972) The Eighteenth Brumaire of Louis Bonaparte.

Marx, K. (1867) Capital Volume 1. On line version. http://www.marxists.org/. archive/ [1 April 2005].

Mason, J. (2002) Qualitative Researching. London: Sage.

Mason, C.N. (2009) Race, Gender and the Recession. New York: NYU Wagner.

Mason, P. (2015) Post-Capitalism. London: Allen Lane.

Mason, P. (2017) The end of capitalism has begun. Guardian 17th July 2015.

McAdam, D., J.D. McCarthy, et al. (1996) Comparative Perspectives on Social Movements. Cambridge: Cambridge University Press.

McCauley, D.A., Heffron, R.J., Stephan, H., and Jenkins, K. (2013) Advancing energy justice: the triumvirate of tenets. International Energy Law Review, 32: 107–110.

McCulloch, J., and Tweedale, G. (2008) Defending the Indefensible: The Global Asbestos Industry and Its Fight for Survival. Oxford, UK: OUPss.

Mcdonald, B.C., Joost, A., De Gouw, J.B. et al. (2018) Volatile chemical products emerging as largest petrochemical source of urban organic emissions. Science, 359(6377): 760–764.

Mcgarvey, D. (2017) Poverty Safari. London: Picador.

McIvor, A. (2012) Germs at work: Establishing tuberculosis as an occupational disease in Britain, c1900–1951. Social History of Medicine, 25: 812–829.

McIvor, A. (2016) Was occupational health and safety a strike issue? Workers, unions and the body in twentieth-century Scotland. Journal of Irish and Scottish Studies, 8(1): 1–45.

McIvor, A., and Johnston, R. (2007) Miners' Lung. Aldershot. Ronald Johnston Books.

McKenna, C., and Olswang, N. (2018) Environmental Permit Refusal: Reversed on Appeal. https://www.lexology.com/library/detail.aspx?g=04617e81-54aa-4436-bca8-a239816f9a09.

Mckenzie, L. (2015) Getting By: Estates, Class and Culture in Austerity Britain. Bristol: Policy Press.

Mckenzie, L. (2017) 'It's not ideal': Reconsidering 'anger' and 'apathy' in the Brexit vote among an invisible working-class. Competition and Change, 21(3): 199–210.

McKibben, B. (2017) Stop talking right now about the threat of climate change. It's here; it's happening, Monday, 11 September 2017. https://www.theguardian.com/commentisfree/2017/sep/11/threat-climate-changehurricane-harvey-irma-droughts.

Meadows, D.H., Meadows, D.L., Randers, J., and Behrens III, W.W. (1972) The Limits to Growth. New York: Universe Books.

Meadows, D.H., Randers, J., and Meadows, D. (2004) Limits to Growth: The 30-Year Update. White River Junction, VT: Chelsea Green Publishing.

Melamed, C.R., and Hartwig, E.A. (2011) Jobs, growth and poverty: what do we know, what don't we know, what should we know? ODI Background Notes. May 2011.

Mercier, L. (2001) Anaconda: Labor, Community, and Culture in Montana's Smelter City. Chicago: University of Illinois Press.

Merritt, A., and Stubbs, T. (2012) Incentives to promote green citizenship in UK Transition Towns. Development, 55(1): 96–103.

Meyer, A. (2000) Contraction and Convergence: The Global Solution to Climate Change. Cambridge: Green Books.

MHCLG. (2018) Empty Housing (England). Ministry of Housing, Communities and Local Government in House of Commons Briefing Paper https://researchbriefings.files.parliament.uk/documents/SN03012/SN03012.pdf.

Midgley, A. (2017) Unite: Candidate Selections Give Us A Chance To Ensure More Working-class Voices Are Heard in Westminster, Unite, 19th October 2017.

Milburn, A. (2017) The government is unable to commit to the social mobility challenge, The Guardian, 2nd December 2017.

Miles, M.B., and Huberman, A.M. (1994) Qualitative Data Analysis: An Expanded Sourcebook (2nd ed.). London: Sage.

Miller, M. (1995) The Third World in Global Environmental Politics. Buckingham: Open University Press.

Miller, C.A., Iles, A., and Jones, C.F. (2013) The Social Dimensions of Energy Transitions. Scientific Culture, 22: 135–148.

Mills, C. (2010) Regulating Health and Safety in the British Mining Industries, 1800–1914. Abingdon: Routledge.

Milojevic, A., Niedzwiedz, C.L., Pearce, J. et al. (2017) Socioeconomic and urban-rural differentials in exposure to air pollution and mortality burden in England. Environmental Health 16(1): 104.

Minnite, L.C., and Fox Piven, F. (2016) Competing Concepts of Class. In M. Shaw and M. Mayo (eds.), Class, Inequality and Community Development. Bristol: Policy Press.

Mitchell, G., and Dorling, D. (2003) An environmental justice analysis of British air quality. Environmental Planning A, 35: 909–929.

Mitchell, R., and Popham, F. (2008) Effect of exposure to natural environment on health inequalities: an observational population study. Lancet 372: 1655–1660.

Mitchell, G., Norman, P., and Mullin, K. (2015) Who benefits from environmental policy? An environmental justice analysis of air quality change in Britain, 2001–2011. Environmental Research Letters, 10(10): 105009.

MOE Norway. (2017) Environmental Information Act. https://www.regjeringen.no/en/dokumenter/environmental-information-act/id173247/.

Monbiot, G. (2017a) How Labour could lead the global economy out of the 20th century, 11th October 2017. https://www.theguardian.com/commentisfree/2017/oct/11/labour-global-economy-planet.

Monbiot, G. (2017b) A lesson from Hurricane Irma: capitalism can't save the planet – it can only destroy it, Wednesday, 13 September 2017. https://www.theguardian.com/commentisfree/2017/sep/13/hurricane-irmacapitalism-growth-economics-environment-financial-crisis.

Mooney, C. (2014) The Science of Why Cops Shoot Young Black Men. Mother Jones. https://www.motherjones.com/politics/2014/12/science-of-racism-prejudice/.

Moore, J.W. (2015) Capitalism in the Web of Life: Ecology and the Accumulation of Capital. London: Verso.

Moore, J. (2017) The case of Lavinia Woodward exposes the troubling inequality at the heart of our justice system, The Independent, 27th May.

Morgan, A. (2012) Inclusive place-based education for 'Just Sustainability'. International Journal of Inclusive Education 16: 627–642.

Morin, R. (2015) Exploring Racial Bias Among Biracial and Single Race Adults: The IAT. Pew Research Centre.

Morss, A. (2019) Pupils have 'human right' to strike for climate, The Ecologist, 15th February 2019.

Mortimer, J. (2013) Think the Greens are the party of the middle-class? Think again. http://liberalconspiracy.org/2013/06/03/think-the-greensare-the-party-of-the-middle-class-think-again/.

Mottus, R., Gale, C., Starr, J., and Deary, I. (2012) 'On the street where you live' Neighbourhood deprivation and quality of life among community-dwelling older people in Edinburgh, Scotland, Social Science and Medicine, 74: 1368–1374.

MPD. (2016) Ministerio de Planificación del Desarrollo. Economic and social development plan 2016–2020 http://www.planificacion.gob.bo/uploads/PDES_INGLES.pdf.

Murphy, J. (2006) Blair: target the problem families to reduce crime. Evening Standard, 23 June.

Namdeo, A., and Stringer, C. (2008) Investigating the relationship between air pollution, health and social deprivation in Leeds, UK. Environment International, 34: 585–591.

Nammo. (2012) We have Improved our Lead-Free Ammunition. www.nammo.com/News/We-have-improved-our-lead-freeammunition-/.

National Audit Office – UK. (2018) The packaging recycling obligations. https://www.nao.org.uk/report/thepackaging-recycling-obligations/.

NEA. (2017) Bridging the Gap: Addressing the cost of living facing UK households this winter, National Energy Action. https://www.nea.org.uk/wp-content/uploads/2017/11/Bridging-the-Gap-NEA.pdf.

NEF. (2017) National Energy Foundation. http://www.nef.org.uk/knowledge-hub/energy-in-the-home/fuel-poverty.

Newell, P. (2005) Race, Class and the Global Politics of Environmental Inequality. Global Environmental Politics, 5(3): 70–94.

Newman, K. (2016) Reconciling Participation and Power in International development: A Case Study. In M. Shaw and M. Mayo (eds.), Class, Inequality and Community Development. Bristol: Policy Press.

O'Leary, N.C., and Sloane, P.J. (2005) The Return to a University Education in Great Britain. National Institute Economic Review, 193(1): 75–89.

O'Shea, B. (2017) Empty Chair: It's Time for Workers on Boards, CLASS. http://classonline.org.uk/docs/Empty_Chair_-_Its_Time_for_Workers_on_Boards_2017_Final.pdf.

ODPM (Office of the Deputy Prime Minister). (2004) The English Indices of Deprivation (revised), London: ODPM.

OECD. (2018a) A Broken Social Elevator? How to Promote Social Mobility, Paris: OECD Publishing.

OECD. (2018b) The Role and Design of Net Wealth Taxes in the OECD. In series: OECD Tax Policy Studies. http://www.oecd.org/ctp/the-role-and-design-of-net-wealth-taxes-in-the-oecd-9789264290303-en.htm.

Oesch, D. (2013) Occupational change in Europe: How Technology and Education Transform the Job structure. Oxford: Oxford University Press.

Ofgem. (2012) https://www.ofgem.gov.uk/ofgem-publications/59153/eradicating-fuel-poverty-and-protecting-vulnerable-customers-nov-2012-pdf.

Ofgem. (2017a) Understanding the profits of the large energy suppliers. https://www.ofgem.gov.uk/gas/retail-market/retail-market-monitoring/understanding-profits-large-energy-suppliers.

Ofgem. (2017b) State of the energy market. https://www.ofgem.gov.uk/system/files/docs/2017/10/state_of_the_market_report_2017_web_1.pdf.

Ogden, P., and Minton, K. (2000) Sensorimotor psychotherapy: one method for processing traumatic memory. Traumatology, I(3): article 3.

Oishi, S., Kesebir, S., and Diener, E. (2011) Income inequality and happiness. Psychological Science, 22(9): 1095–1100.

ONS. (2017) Office for National Statistics. https://www.ons.gov.uk/peoplepopulationandcommunity/birthsdeathsandmarriages/deaths/bulletins/excesswintermortalityinenglandandwales/2015to2016provisionaland2014to2015final.

Orr, D. (2017). https://www.theguardian.com/commentisfree/2017/jul/24/humans-losing-touch-nature-alcoholism-gambling?CMP=share_btn_fb. Monday. 24 July 2017.

Ossowski, S. (1963) Class Structure in the Social Consciousness. London: Routledge and Kegan Paul.

Oxfam. (2013) The cost of inequality: how wealth and income extremes hurt us all. https://www.oxfam.org/sites/www.oxfam.org/files/cost-of-inequalityoxfam-mb180113.pdf.

Oxfam. (2014) Flood Risk.

Oxfam. (2015) Extreme Carbon Inequality, Why the Paris climate deal must put the poorest, lowest emitting and most vulnerable people first. https://www.oxfam.org/sites/www.oxfam.org/files/file_attachments/mb-extremecarbon-inequality-021215-en.pdf.

Oxfam. (2016) An Economy For the 1%: How privilege and power in the economy drive extreme inequality and how this can be stopped. http://policypractice.oxfam.org.uk/publications/an-economy-for-the-1-howprivilege-and-power-in-the-economy-drive-extreme-inequ-592643.

Pacheco-Vega, R., and Parizeau, K. (2018) Doubly Engaged Ethnography: Opportunities and Challenges When Working With Vulnerable Communities. International Journal of Qualitative Methods, 17(1): 1–13.

Paehlke, R. (1989) Environmentalism and the Future of Progressive Politics, New Haven, CT: Yale University Press.

Pakulski, J., and Waters, M. (1996) The Death of Class. London: Sage.

Pandit, R., and Laband, D.N. (2009) Economic well-being, the distribution of income and species imperilment. Biodiversity and Conservation, 18: 3219–3233.

Peace, J., and Milne, E.-J. (2010) Participation and community on Bradford's traditionally white estates. York: JRF.

Pearce, J.R., Richardson, E.A., Mitchell, R.A., and Shortt, N.K. (2010) Environmental justice and health: the implications of the socio-spatial distribution of multiple environmental deprivation for health inequalities in the United Kingdom. Transactions of the Institute of British Geographers, 35(4): 522–539.

Pellegriti, G., et al. (2013) Worldwide Increasing Incidence of Thyroid Cancer: Update on Epidemiology and Risk Factors. Journal of Cancer Epidemiology, 2013, Article ID 965212, 10 pages.

Pellow, D.N. (2007) Resisting Global Toxics: Transnational Movements for Environmental Justice. Cambridge, MA: The MIT Press.

Pellow, D.N. (2018) What is Critical Environmental Justice? Cambridge: Polity.

Pellow, D., and Brulle, R. (eds.). (2005) Power, Justice and the Environment: A Critical Appraisal of the Environmental Justice Movement, Cambridge, MA: MIT Press.

Pickett, K., and Wilkinson, R. (2010) The Spirit Level: Why Greater Equality Makes Societies Stronger. New York: Bloomsbury Publishing.

Piercy, K. (2016) Classist innuendo about educated Remain voters and the 'white van men' of Leave has revealed something very distasteful about Britain. The Independent, 20th June 2016.

Piff, P.K., Stancato, D.M., Côté, S., Mendoza-Denton, R., and Keltner, D. (2012) Higher social class predicts increased unethical behavior. February 27, 2012. Proceedings National Academy Sciences, USA.

Piketty, T. (2014) Capital in the Twenty-First Century. Cambridge, MA: Harvard University Press.

Porges, S. (2004) Neuroception: a subconscious system for detecting threats and safety. Zero to three, May 2004, 19–24.

Porritt, J. (2005, revised 2007) Capitalism: as if the World Matters, London: Earthscan.

Price Waterhouse Cooper. (2018) UK Economic Outlook 2018. https://www. pwc.co.uk/services/economics-policy/insights/uk-economic-outlook.html.

Pulido, L. (1998) Development of the 'People of Color' Identity in the Environmental Justice Movement of the Southwestern United States. Socialist Review, 26(3–4): 145–180.

Pye, S.K., King, K., and Sturman, J. (2006) Air Quality and Social Deprivation in the UK: An Environmental Inequalities Analysis-Final Report to Defra, Contract rmp/2035. https://uk-air.defra.gov.uk/assets/documents/reports/cat09/0701110944_AQinequalitiesFNL_AEAT_0506.pdf.

Raaschou-Nielsen, O., et al. (2013) Long-term exposure to traffic-related air pollution and diabetes-associated mortality: a cohort study. Diabetologia, 56(1): 36–46.

Raftery, A.E. et al. (2017) Less than 2°C warming by 2100 unlikely. Nature Climate Change, 7: 637–641.

Ranft, U., et al. (2009) Long-term exposure to traffic-related particulate matter impairs cognitive function in the elderly. Environmental Research, 109(8): 1004–1011.

Räthzel, N., and Uzzell, D. (2011) Trade unions and climate change: The jobs versus environment dilemma. Global Environmental Change-Human and Policy Dimensions, 21(4): 1215–1223.

Räthzel, N., and Uzzell, D. (2012) Mending the breach between labour and nature: Environmental engagements of trade unions and the North-South divide. Interface, 4(2): 81–100.

Räthzel, N., and Uzzell, D. (eds.). (2013) Trade Unions in the Green Economy: Working for the Environment. London: Routledge.

Räthzel, N., Uzzell, D., and Elliott, D. (2010) Can Trade Unions Become Environmental Innovators? Soundings, 46: 76–87.

Raworth, K. (2017) Doughnut Economics: Seven Ways to Think Like a 21st-Century Economist. White River Junction, VT: Chelsea Green Publishing.

Reed, M.G., and George, C. (2011) Where in the world is environmental justice? Progress in Human Geography 35(6): 835–842.

Resolution Foundation. (2017) 21st Century Britain has seen a 30 per cent increase in second home ownership. http://www.resolutionfoundation.org/media/press-releases/21st-century-britain-has-seen-a-30-per-cent-increasein-second-home-ownership/.

Resolution Foundation. (2019) https://www.resolutionfoundation.org/theme/jobs-skills-and-pay/.

Revkin, A. (2004) The Burning Season: The Murder of Chico Mendes and the Fight for the Amazon Rain Forest. Washington: First Island Press, Sherewater Books Printing.

Richardson, H. (2017) Oxbridge uncovered: More elitist than we thought By BBC News education reporter. BBC, 20th October 2017. http://www.bbc.co.uk/news/education-41664459.

Richardson, E., Shortt, N., and Mitchell, R. (2010) The mechanism behind environmental inequality in Scotland: which came first, the deprivation or the landfill? Environment and Planning A, 42: 223–240.

Rifkin, J. (2010) The Empathic Civilization. https://www.ted.com/talks/jeremy_rifkin_on_the_empathic_civilization.

Rivera, L.A. (2015) Pedigree: How Elite Students Get Elite Jobs. Princeton, NJ: Princeton University Press.

Rivera, L.A., and Tilcsik, A. (2016) Class Advantage, Commitment Penalty: The Gendered Effect of Social Class Signals in an Elite Labor Market. American Sociological Review, 81(6): 1097–1131.

Roberts, K., Cavill, N., Hancock, C., and Rutter, H. (2013) Social and Economic Inequalities in Diet and Physical Activity. London: Public Health England.

Roll for the Soul. (2017) There's no easy way to say it: we're closing. http://www.rollforthesoul.org/theres-no-easy-way-to-say-it-were-closing.

Rose-Ackerman, S., and Palifka, B.J. (2016) Corruption and government: Causes, consequences, and reform, 2nd edition. New York: Cambridge University Press.

Roy, J.-P. (2004) Socioeconomic status and health: a neurobiological perspective. Medical Hypotheses, 62(2): 222–227.

Royal College of Psychiatrists. (2010) No health without Royal College of Psychiatrists Position statement PS4/2010 public mental health the case for action. http://www.rcpsych.ac.uk/pdf/Position%20Statement%204%20website.pdf.

RRF. (2002) Household Waste Behaviour in London. Skipton, UK: Resource Recovery Forum.

RRF. (2004) Household Waste Behaviour in London, Phase 2: High, medium and low recyclers: attitudes, behaviours and needs. Skipton, UK: Resource Recovery Forum.

RSPB. (2010) Every Child Outdoors. In Syal, R. One in eight workers struggle to afford food, finds TUC survey. https://www.theguardian.com/society/2017/sep/07/one-in-eight-workers-struggle-to-afford-food-findstuc-survey?CMP=share_btn_tw, Thursday 7 September 2017.

Rubin, M., Denson, N., Kilpatrick, S., Matthews, K.E., Stehlik, T., and Zyngier, D. (2014) "I Am Working-Class": Subjective Self-Definition as a Missing Measure of Social Class and Socioeconomic Status in Higher Education Research. Educational Researcher, 43: 196–200.

Ruiz, S. (2018) Grenfell United in Booth, R. Grenfell Tower: fire-resistant cladding plan was dropped. The Guardian, 8th May.

Sanders, B. (2009) The Green Zone: The Environmental Costs of Militarism. Oakland, CA: AK Press.

Sandler, R., and Pezzullo, P.C. (eds.). (2007) Environmental Justice and Environmentalism: The Social Justice Challenge to the Environmental Movement. Cambridge, MA: Massachusetts Institute of Technology, pp. 105–135.

Sapolsky, R. (2018) The science of inequality. Scientific American. November 2018.

Sarkar, S. (1999) Eco-Socialism or Eco-Capitalism: A critical Analysis of Humanity's Fundamental Choices. London: Zed Books.

Saunders, P. (2007) A systematic review of the evidence of an increased risk of adverse birth outcomes in populations living in the vicinity of landfill waste disposal sites. In Population Health and Waste Management: Scientific Data and Policy Options. Report of a WHO Workshop Rome, Italy, 29–30 March 2007; World Health Organization: Geneva, Switzerland.

Savage, M. (2000) Class analysis and social transformation. Milton Keynes: Open University Press.

Savage, M., et al. (2013) A new model of social class? Findings from the BBC's great British class survey experiment. Sociology, 47(2): 219–250.

Savage, M., et al. (2015) Social class in the 21st century. London: Penguin.

Schaefer, T., and LaGrange, A. (2018) Measure to combat our throwaway tech culture would be win-win for repair workers and landfills, StarTribune, MARCH 2, 2018. http://www.startribune.com/measure-tocombat-our-throwaway-tech-culture-would-be-win-win-for-repair-workersand-landfills/475703233/.

Schandl, H., Hatfield-Dodds, S., Wiedmann, T., et al. (2016) Decoupling global environmental pressure and economic growth: Scenarios for energy use, materials use and carbon emissions. Journal of Cleaner Production, 132: 45–56.

Schmelzer, M., and Passadakis, A. (2011) Postwachstum. Hamburg: VSA Verlag.

Schnaiberg, A., and Gould, K. (2000) Environment and Society: The Enduring Conflict. New York: St Martin's Press.

Schumacher. (2017) https://www.eventbrite.co.uk/e/all-systems-go-tickets-36018935606.

Scottish Government. (2017) The Socio-Economic Duty: A Consultation. http://www.gov.scot/Resource/0052/00522478.pdf.

SDKP. (2017) Transforming our world: the 2030 Agenda for Sustainable Development, Sustainable Development Knowledge Platform. https://sustainabledevelopment.un.org/post2015/transformingourworld.

SDRN. (2004) Environment and Social Justice Review: Rapid Research and Evidence Review. Revised Version, December 2004. London: Policy Studies Institute.

SEER. (2013) The Surveillance, Epidemiology, and End Results (SEER) Program of the National Cancer Institute (NCI) 'Cancer Statistics'. www.seer.cancer.gov.

Seery, E. (2014) Working for the Many. Oxford: Oxfam.

Seligman, M.E.P. (1972) Learned helplessness. Annual Review of Medicine, 23(1): 407–412.

Sennett, R., and Cobb, J. (1973) The Hidden Injuries of Class. New York: NY: Vintage Books.

Seyfang, G. (2009) Report of the 2009 Membership Survey, Norwich Transition Town.

Seyfang, D., and A. Haxeltine. (2012) Growing grassroots innovations: exploring the role of community-based initiatives in governing sustainable energy transitions. Environmental Planning C, 30(3): 381–400.

Shah, P.S., and Balkhair, T. (2011) Air pollution and birth outcomes: a systematic review. Environment International, 37(2): 498–516.

Shaw, M., and Mayo, M. (eds.). (2016) Class Inequality and Community Development. Bristol: Policy Press.

Shelter. (2013) Sharp Rise in Number of Homeless Families. http://england.shelter.org.uk/news.

Shelter. (2017) More than 300,000 people in Britain homeless today. https://england.shelter.org.uk/media/press_releases/articles/more_than_300,000_people_in_britain_homeless_today.

Shildrick, T., MacDonald, R., Webster, C., and Garthwaite, K. (2012) Poverty and insecurity: Life in low pay, no pay Britain. Bristol: Policy Press.

Siegler, V., et al. (2010) Social inequalities in fatal accidents and assaults: England and Wales, 2001–03, Office for National Statistics, Health Statistics Quarterly 48, Winter 2010.

Silva, E. (2015) Class in contemporary Britain: comparing the Cultural Capital and Social Exclusion (CCSE) project and the Great British Class Survey (GBCS). The Sociological Review, 63(2): 373–392.

Silverman, R.A., Ito, K., et al. (2010) Association of Ambient Fine Particles With Out-of-Hospital Cardiac Arrests in New York City. American Journal of Epidemiology 172(8): 917–923.

Simons, A.M.W. et al. (2017) Perceived classism and its relation with socioeconomic status, health, health behaviours and perceived inferiority: the Dutch Longitudinal Internet Studies for the Social Sciences (LISS) panel, International Journal of Public Health, 62(4): 433–440.

SIPRI and CAAT. (2016) Special Treatment UK Government Support for the Arms Industry and Trade. https://www.caat.org.uk/resources/publications/economics/special-treatment.pdf.

Skeggs, B. (1997) Formations of Class and Gender. London: Sage.

Skeggs, B. (2004) Class, Culture, Self. London: Routledge.

Slater, T. (2011) Gentrification of the City. In G. Bridge and S. Watson (eds.), The New Companion to the City. Oxford: Blackwell, pp. 571–585.

Smith, D.E. (1987) The Everyday World as Problematic: A Feminist Sociology. Milton Keynes: Open University Press.

Smith, A. (2011) The transition town network: a review of current evolutions and renaissance. Social Movement Studies, 10(1): 99–105.

Smith, L., Mao, A., and Deshpande, A. (2015) "Talking Across Worlds": Classist Microaggressions and Higher Education, Journal of Poverty, 20(2): 127–151.

Social Mobility Commission. (2017) Social mobility policies between 1997 and 2017: time for change.

South Wales Argus. (2010) Welsh secretary sees Magor brewery's green efforts. http://www.southwalesargus.co.uk/news/6199794.display/.

Sovacool, B.K., and Dworkin, M.H. (2015) Energy justice: conceptual insights and practical applications, Applied Energy, 142: 435–444.

Standing, G. (2011) The Precariat: The New Dangerous Class. London: Bloomsbury Academic.

Staub, C. (2018) EPA: Plastics Recycling Rate Declines, Plastics Recycling Update. https://resource-recycling.com/plastics/2018/08/01/epa-u-s-plasticsrecycling-rate-declines/.

Steffen, W., et al. (2015) Planetary Boundaries: Guiding human development on a changing planet. Science, 347(6223): 1259855.

Stephens, C., et al. (2001) Environmental justice: Rights and means to a healthy environment for all. https://friendsoftheearth.uk/sites/default/files/downloads/environmental_justice.pdf.

Stevenson, N. (2012) Localization as subpolitics: the transition movement and cultural citizenship. International Journal of Cultural Studies, 15(1): 65–79.

Stevis, D. (2013) Green Jobs? Good Jobs? Just Jobs? USA Labour Unions Confront Climate Change. In N. Räthzel and D. Uzzell (eds.), Trade Unions in the Green Economy: Working for the Environment. London: Routledge.

Stevis, D., and Felli, R. (2015) Global labour unions and just transition to a green economy, International Environmental Agreements: Politics, Law and Economics, 15(1): 29–43.

Stiglitz, J.E. et al. (2010) Report by the Commission on the. Measurement of Economic. Performance and Social Progress. Professor, Chair, Columbia University. http://ec.europa.eu/eurostat/documents/118025/118123/Fitousi+Commission+report.

Stiglitz, J.E., Stern, N., Duan, M., et al. (2018) Report of the High-Level Commission on Carbon Prices. Carbon Pricing Leadership Coalition.

Stockton, H., and Campbell, R. (2011) Time to Reconsider UK Energy and Fuel Poverty Policies? York: Joseph Rowntree Foundation.

Stuart, G. et al. (2014) House of Commons Education Committee, Underachievement in education by white working class children.

Sue, D.W. (2010) Microaggressions in. Everyday Life: Race, Gender, and Sexual. Orientation. NJ: John Wiley & Sons.

Sustainable Development Commission. (2010) Improving Young People's Lives: The role of the environment in building resilience, responsibility and employment chances.

Sutton Trust. (2009) The Educational Backgrounds of Leading Lawyers, Journalists, Vice Chancellors, Politicians, Medics And Chief Executives. https://www.suttontrust.com/wp-content/uploads/2009/04/ST_MilburnSubmission-1.pdf.

Sweeney, S. (2012) U.S. Trade Unions and the Challenge of 'Extreme Energy' The Case of the TransCanada Keystone XL Pipeline. In N. Räthzel and D. Uzzell (eds.), Trade Unions in the Green Economy. Working for the Environment. Oxford/New York: Earthscan/Routledge.

Taylor, A., and Loopstra, R. (2016) Too Poor to Eat: Food insecurity in the UK. London: Food Foundation.

Taylor, M., and Wilson, M. (2016) Community organising for social change: The scope for class politics. In M. Shaw and M. Mayo (eds.), Class, inequality and community development. Bristol: Policy Press.

Telegraph. (2015) Council waste tens of thousands on 'green' beech tree art project, 1st October 2015. http://www.telegraph.co.uk/news/uknews/11905612/Council-waste-tens-of-thousands-on-green-beech-tree-artproject.html.

The Committee on Climate Change. (2017) Energy Prices and Bills: Impact of meeting carbon budgets. London. https://www.theccc.org.uk/wp-content/uploads/2017/03/Energy-Prices-and-Bills-Committee-on-Climate-Change-March-2017.pdf.

The Equality Trust. (2017) https://www.equalitytrust.org.uk/.

'The Lancet Commission on Pollution and Health'. (2017) http://www.thelancet.com/journals/lancet/article/PIIS0140-6736(17)32345-0/fulltext and http://www.thelancet.com/commissions/pollution-and-health.

The Localism Act. (2011) In: Public General Acts – Elizabeth II. London: The Stationery Office, chapter 20.

The Open University. (2008) The Open University household waste study: Key findings for 2007, London: DEFRA.

The UK Low Carbon Industrial Strategy, www.gov.uk.

The UK Low Carbon Transition Plan: National Strategy for Climate & Energy. http://webarchive.nationalarchives.gov.uk/20100512172800/http://www.decc.gov.uk/en/content/cms/publications/lc_trans_plan/lc_trans_plan.aspx.

The Women's Trade Union League (WTUL). (1899) Potteries Fund Register Lead poisoning in the potteries, 1899, TUC Library Collections, London Metropolitan University.

Thomson, H., Thomas, S., Sellström, E., and Petticrew, M. (2013) Housing improvements for health and associated socio-economic outcomes: a systematic review. Campbell Systematic Reviews, No. 2.

Titheridge, H., Christie, N., Mackett, R., Hernández, D.-O., and Ye, R. (2014) Transport and Poverty: A review of the evidence. London: UCL.

Torras, M., & Boyce, J.K. (1998) Income, inequality, and pollution: a reassessment of the environmental Kuznet Curve. Ecological Economics, 25: 147–160.

Torras, M., S.A. Moskalev, J.K. Hazy, and A.S. Ashley. (2011) An econometric analysis of ecological footprint determinants: implications for sustainability, International Journal of Sustainable Society, 3: 258–275.

Townshend, T.G. (2017) Toxic high streets. Journal of Urban Design, 22(2): 67–186.

Transition Network. (2017) https://transitionnetwork.org/about-themovement/.

TUC. (2008) A green and fair future. For a just transition to a low carbon economy. Touchstone Pamphlet no 3. Trades Unions Congress: London.

TUC. (2010) Swords of justice and civic pillars: The case for greater engagement between British trade unions and community organisations. London: TUC.

TUC. (2017) 1 in 8 workers are skipping meals to make ends meet – we need action on pay now. https://www.tuc.org.uk/blogs/1-8-workers-are-skipping-meals-make-ends-meet-%E2%80%93-we-need-action-pay-now.

Tweedale, G. (2000) Magic Mineral to Killer Dust: Turner and Newall and the Asbestos Hazard. Oxford.

UCC. (1987) Toxic Wastes and Race in the United States: A National Report on the Racial and Socio-Economic Characteristics of Communities with Hazardous Waste Sites New York, United Church of Christ Commission for Racial Justice.

UK Gov. (2018) Table 615: vacant dwellings by local authority district: England, from 2004. https://www.gov.uk/government/statistical-data-sets/live-tableson-dwelling-stock-including-vacants.

UK Working Group on the Primary Prevention of Breast Cancer. (2013) Breast Cancer an Environmental Disease. https://tippingpointnorthsouth.files.wordpress.com/2013/06/breast-cancer-summary.pdf.

Ulrich, R.S. (1984) View through window may influence recovery from surgery. Science 224: 420–421.

UNCED. (1992) The Rio Declaration on Environment and Development. http://www.unesco.org/education/nfsunesco/pdf/RIO_E.PDFUNEP.

UNCSD. (2012) The Future we Want, Outcome document of the United Nations Conference on Sustainable Development, Rio de Janeiro, Brazil, 20–22 June 2012. https://sustainabledevelopment.un.org/content/documents/733FutureWeWant.pdf.

UNECE. (1999) Aarhus Convention, United Nations Economic Commission for Europe.

UNEP. (2006) Final Resolution of the Trade Union Assembly on Labour and the Environment, Nairobi, Kenya, 15–17 January, 2006.

UNEP. (2008) Green Jobs: Towards Decent Work in a Sustainable, Low-Carbon World. United Nations Environment Programme. https://digitalcommons.ilr.cornell.edu/cgi/viewcontent.cgi?referer=https://www.google.co.uk/&http sredir=1&article=1057&context=intl.

UNEP. (2011) Towards a Green Economy. Pathways to Sustainable Development and Poverty Reduction. Nairobi, Kenya: UNEP.

UNEP. (2017) United Nations Environment Programme Twitter, 21st February 2017. https://twitter.com/UNEP/status/834116292504330242.

UNFCCC. (2010) Decision 1/CP.16. The Cancun Agreements: Outcome of the work of the Ad Hoc Working Group on Long-term Cooperative Action under the Convention. http://unfccc.int/resource/docs/2010/cop16/eng/07a01.pdf.

United Nations Human Development Report 2015. (2015) New York: United Nations Development Programme.

US GAO (General Accounting Office). (1983) Siting of Hazardous Waste Landfills and Their Correlation with Racial and Economic Status of Surrounding Communities, GAO/RCED-83–168, Washington, DC: Government Printing Office.

Varoufakis, Y. (2017) Bristol, Bristol Festival of Ideas 20th October, 2017.

Vaughan, A. (2017) What's our cut? Energy firm lets customers see profit margins, The Guardian, 8th May 2017 https://www.theguardian.com/business/2017/may/08/octopus-energy-tariff-customers-see-profitmargins-gas-electricity.

Veblen, T. (1899) The Theory of the Leisure Class: An Economic Study in the Evolution of Institutions. New York: The Macmillan Company.

Verbist, G., M. Förster, and M. Vaalavuo. (2012) The Impact of Publicly Provided Services on the Distribution of Resources: Review of New Results and Methods. OECD Social, Employment and Migration Working Papers, No. 30, OECD Publishing.

Vona, F., and F. Patriarca. (2011) Income inequality and the development of environmental technologies. Ecological Economics, 70: 2201–2213.

Vosper, N. (2016) What makes me tired when organising with middle-class comrades, Guardian 8th June 2016.

Vrijheid, M. (2000) Health effects of residence near hazardous waste landfill sites: A review of epidemiologic literature. Environmental Health Perspectives, 1081: 101–112.

Vucetich, J.A. (2017) Are humans and nature fundamentally one and the same? Center for Humans and Nature, 18 Aug 2017.

Walker, G. (2012) Environmental Justice: Concepts, Evidence and Politics, London: Routledge.

Walker, J. (2017) MP Ian Lavery says he'd rather 'die in a gutter' than lose his working-class accent, Chronicle Live 2nd Oct 2017. http://www.chronicle-live.co.uk/news/north-east-news/mp-ian-lavery-says-hed-13705354.

Walker, G., and Burningham, K. (2011) Flood risk, vulnerability and environmental justice: evidence and evaluation of inequality in a UK context. https://core.ac.uk/download/pdf/9550048.pdf.

Walker, G., Mitchell, G., Fairburn, J., and Smith, G. (2003a) Environmental Quality and Social Deprivation. Phase II: National Analysis of Flood Hazard,

IPC Industries and Air Quality (Bristol: The Environment Agency), p 133. R&D Project Record E2-067/1/PR1.

Walker, G., Fairburn, J., Smith, G., and Mitchell, G. (2003b) Deprived Communities Experience Disproportionate Levels of Environmental Threat. Bristol: Environment Agency.

Walker, G., Mitchell, G., Fairburn, J., and Smith, G. (2005a) Industrial Pollution and Social Deprivation: Evidence and Complexity in Evaluating and Responding to Environmental Inequality. Local Environment, 10(4): 361–377l.

Walker, G., Gordon, M., Fairburn, J., and Smith, G. (2005b) Environmental Quality and Social Deprivation Phase II: National Analysis of Flood Hazard, IPC Industries and Air Quality; Environment Agency: Bristol, UK, 2005.

Walker, G., Burningham, K., Fielding, J., Smith, G., Thrush, D., and Fay, H. (2007) Addressing environmental inequalities: flood risk. Bristol: Environment Agency.

Warwick, H. (2017) What if all students spent a year working the land before university? The Guardian 17 July 2017.

We Own It. (2017) https://weownit.org.uk/.

Webb, S., and Webb, B. (1902) Industrial Democracy, London.

Weber, M. (1958) Class, Status and Party. In Max Weber: Essays in Sociology, edited by H. Gerth and C. Wright Mills. Oxford, UK: Oxford University Press, pp. 180–195.

Weeden, K.A., Kim, Y.M., Di Carlo, M., and Grusky, D.B. (2007) Social class and earnings inequality. American Behavioral Scientist 50(5): 702–736.

Wei, M., Patadia, S., and Kammen, D.M. (2010) Putting renewables and energy efficiency to work. Energy Policy 38(2): 919–931.

Weir, K. (2012) The pain of social rejection. American Psychological Association, 43.

Wheatley, S., Sovacool, B.J., and Sornette, D. (2016) Reassessing the safety of nuclear power. Energy Research and Social Science, 15: 96–100.

Wheeler, B. (2004) Health-related environmental indices and environmental equity in England and Wales. Environment and Planning A, 36(5): 803–822.

White, R. (1995) Are you an Environmentalist or do you Work for a Living? In W. Cronon (ed.), Uncommon Ground: Rethinking the Human Place in Nature. New York: W. W. Norton & Company, pp. 171–185.

Whittle, R., et al. (2010) After the Rain – learning the lessons from flood recovery in Hull, final project report for 'Flood, Vulnerability and Urban Resilience: a real-time study of 36 local recovery following the floods of June 2007 in Hull', Lancaster University, Lancaster UK.

WHO. (2016) Preventing disease through healthy environments. http://apps. who.int/iris/bitstream/10665/204585/1/9789241565196_eng.pdf?ua=1.

WHO. (2018a) Noncommunicable diseases. http://www.who.int/news-room/ fact-sheets/detail/noncommunicable-diseases.

WHO. (2018b) WHO Global Ambient Air Quality Database (update 2018). http://www.who.int/airpollution/data/en/.

WHO and UNICEF. (2017) Progress on drinking water, sanitation and hygiene: 2017 updates and SDG baselines. Geneva: WHO and UNICEF.

Wiedmann, T.O. et al. (2013) The material footprint of nations, Proceedings of the National Academy of Sciences of the United States of America PNAS, Proceedings of the National Academy of Sciences, 112(20): 6271–6276.

Wilkinson, R., and Pickett, K. (2009) The Spirit Level: Why More Equal Societies Almost Always Do Better. London: Allen Lane.

Wilkinson, R., and Pickett, K. (2019) The Inner Level: How More Equal Societies Reduce Stress, Restore Sanity and Improve Everyone's Well-Being. Penguin Press.

Williams, S. (2016) Bristol Green Capital Money Still Under Wraps. https:// stephenwilliamsmp.wordpress.com/2016/03/04/bristol-greencapital- money-still-under-wraps/.

Williams, J.C. (2017) White Working Class: Overcoming Class Cluelessness in America, Harvard Business Review Press.

Windsteiger, L. (2017) How our narrowing social circles create a more unequal world. https://www.theguardian.com/inequality/2017/jun/27/people-likeus- why-narrowing-social-circles-create-more-unequal-world?platform=hootsuite.

Woodward, D., and Simms, A. (2006) Growth isn't Working: The Unbalanced Distribution of Benefits and Costs from Economic Growth. Series: Re-thinking Poverty 1. London: New Economics Foundation.

World Bank. (2017) Data. Washington, DC: World Bank.

World Economic Forum. (2017) Global Risks Report. https://www.weforum. org/reports/the-global-risks-report-2017.

World Inequality Report. (2018) http://wir2018.wid.world/files/download/ wir2018-full-reportenglish.Pdf.

Wright, E.O. (2001) Foundations of Class Analysis: A Marxian Perspective. In Reconfigurations of Class and Gender, edited by J. Baxter and M. Western. Stanford, CA: Stanford University Press, pp. 14–27.

Wright, E.O. (2005) Foundations of a Neo-Marxist Class Analysis. In E.O. Wright, Approaches to Class Analysis, NY: Cambridge University Press, pp. 4–30.

WWF World Wildlife Fund. (2006) Living planet report 2016. Gland, Switzerland: WWF.

Wynes, S., and Nicholas, K.A. (2017) The climate mitigation gap: education and government recommendations miss the most effective individual actions. Environmental Research Letters 12: 074024.

Yandle, B., Vijayaraghavan, M., and Bhattarai, M. (2002) The Environmental Kuznets Curve: A Primer. The Property and Environment Research Center.

Index

© The Author(s) 2020
K. Bell, *Working-Class Environmentalism*,
https://doi.org/10.1007/978-3-030-29519-6

283